Wild Ohio

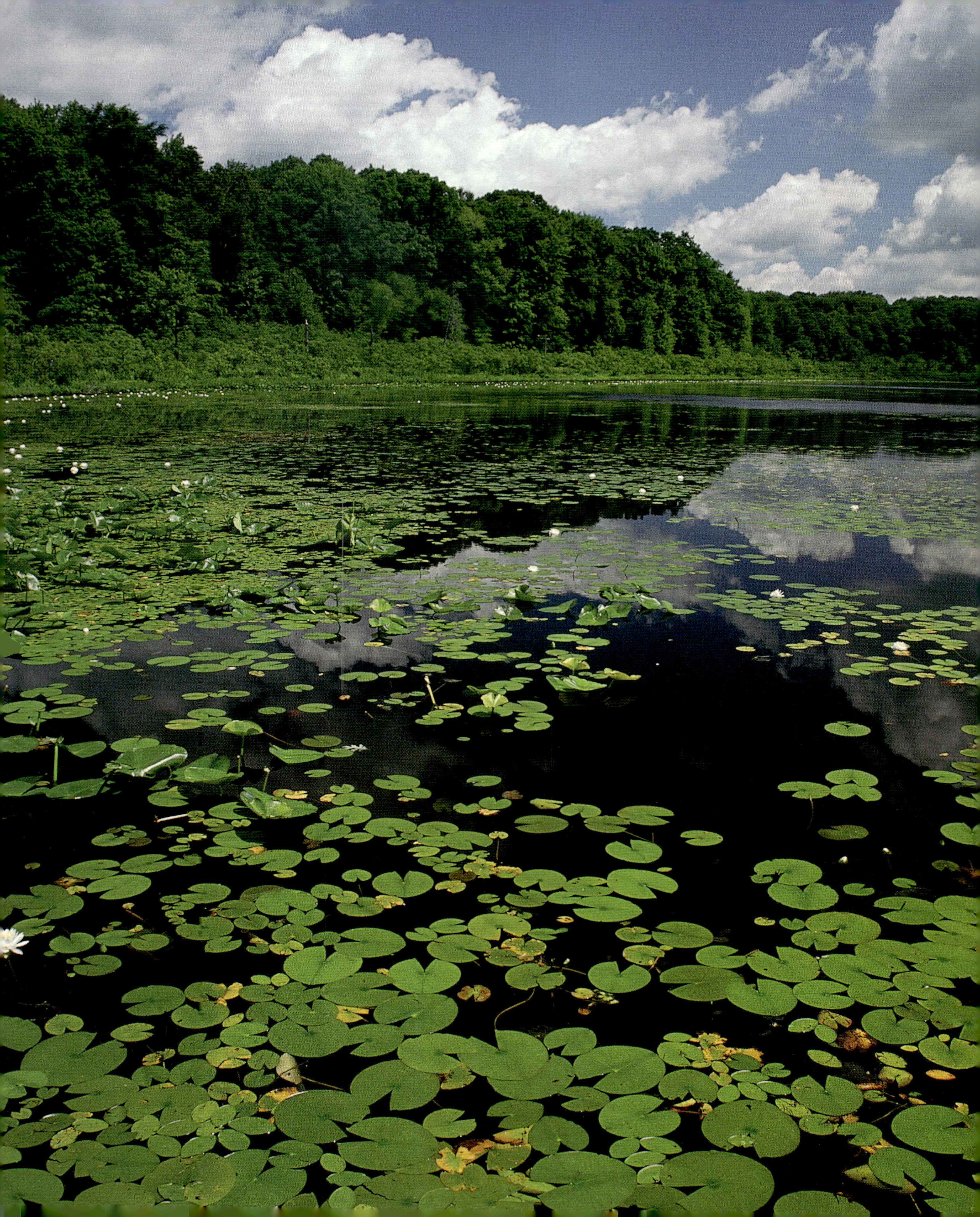

Wild Ohio

The Best of Our Natural Heritage

Jim McCormac
and Gary Meszaros

The Kent State University Press
Kent, Ohio

Library of Congress Catalog Card Number 2008021132
ISBN 978-0-87338-985-3

Manufactured in Korea

Library of Congress Cataloging-in-Publication Data
McCormac, James S., 1962–
Wild Ohio : the best of our natural heritage / Jim McCormac and Gary Meszaros.
p. cm.
Includes bibliographical references and index.
ISBN 978-0-87338-985-3 (alk. paper)∞
1. Natural history—Ohio—Pictorial works.
2. Natural areas—Ohio—Pictorial works.
3. Parks—Ohio—Pictorial works.
4. Wildlife refuges—Ohio—Pictorial works.
5. Forest reserves—Ohio—Pictorial works.
6. Ohio—History, Local—Pictorial works.
7. Natural history—Ohio.
8. Ohio—History, Local.
9. Ohio—Description and travel.
10. Ohio—Environmental conditions.
I. Meszaros, Gary. II. Title.
QH105.O3M38 2008
508.771—dc22 2008021132

British Library Cataloging-in-Publication data are available.

21 20 19 18 17 5 4 3 2

This book is dedicated to all Ohioans, past and present, with an interest in conservation. We especially hope the words and photos within might serve to inspire future conservationists.

Contents

Foreword

I can think of few individuals in Ohio more qualified than Jim McCormac and Gary Meszaros to write and illustrate a book on the best remaining examples of the natural habitats that were present within Ohio a little more than 200 years ago.

Both Gary and Jim have devoted a significant portion of their professional and personal lives to exploring the most pristine remnants of our original forests, glacial lakes (including the largest—Lake Erie), vast marshes, swamps, bogs, fens, oak savannas, rock cliffs, and streams large and small throughout the state. Both have spent countless hours exploring rare habitats throughout Ohio that support the last-known occurrences of particular members of our state's endangered flora and fauna. Their respective reputations for discovering new populations of rarities, both within and outside their respective areas of expertise, are legendary among professional biologists and the private and public managers of the extensive system of parks and nature preserves scattered across our state.

Pairing Jim's encyclopedic knowledge of the natural communities in Ohio's physiographic provinces with Gary's photographic talent has produced a book for everyone who champions the protection of wild Ohio.

What conservationists now consider to be the most important and imperiled remnants of our natural heritage has changed dramatically since the Ohio Natural Heritage Program was established in 1980. This has been due in large part to the efforts of amateur and professional biologists who have discovered more than 100 new, ecologically significant natural areas and contributed thousands of records of new rare plant and animal occurrences in Ohio to the Natural Heritage Program and the Ohio Division of Wildlife. Several of the natural areas that Gary and Jim selected to feature in this book are among those new sites.

Interestingly, more than 170 plant species thought to be rare when the Ohio Natural Heritage Program was established have been removed from the Ohio Rare Plant List due to proof that they are, in fact, not rare. In some cases, this improved knowledge has helped us see that some habitats are much more common, while other habitats thought to be relatively secure are actually threatened.

The thousands of Ohio rare plant and animal occurrence records held, respectively, by the Ohio Natural Heritage Program and Ohio Division of Wildlife are used to determine which habitat types and specific geographic sites are most in need of protection. Efforts during the past few decades to improve our collective knowledge of the distribution and rarity of several biological groups, including moths, butterflies, mosses, lichens, dragonflies, caddis flies, fish, and freshwater mussels, have identified dozens of sites in the state as biologically important habitats that should be protected.

While it is important to continue supporting these efforts, we should also direct funds to help develop the expertise and field support for the study of other biological groups that are still frontiers in our knowledge base, such as ground beetles, grasshoppers, spider wasps, land snails, and mammals. We need to identify and protect the habitats in which these creatures occur before they are lost.

The future residents of Ohio will only be able to witness the outstanding examples of our natural heritage chosen for this book if its current residents devote more resources toward reducing and, ideally, stopping the spread of invasive plants and animals into our natural areas. Without proper management of our parks, wildlife preserves, private nature preserves, and conservation easements, most of the outstanding habitats featured in this book will be irreparably altered over the next fifty years.

The overpopulation of white-tailed deer and invasion of non-native earthworms are particular problems. These two creatures are producing biological deserts beneath many of our forest canopies. There is presently no known method to reverse this trend. Perhaps this book will inspire someone to devise a solution. Fortunately, state, federal, and local governmental agencies, nonprofit conservation organizations, and land trusts across Ohio are beginning to earmark funds for this sort of environmental stewardship.

Global climate change is the newest challenge to our protected lands. The projected warming of our climate will change the composition of dominant species on many of the nature preserves in the state. However, it's important to remember the flexibility inherent in natural systems. If our protected lands receive proper environmental stewardship, most of the native habitats included in this book will still support outstanding biodiversity fifty years from now.

It's my hope that this book will move people to support the statewide effort to steward and protect the best remaining examples of Ohio's natural heritage.

Dr. James K. Bissell
Director of Conservation
Curator of Botany
The Cleveland Museum of Natural History

Preface

We hope this book never becomes an epitaph for what once was. Ohio is incredibly rich in biodiversity, possibly more so than any other midwestern state. But in the course of assembling a documentary book like this one, it's impossible not to think about what we've lost. Some three hundred years ago the first European settlers stepped into the Ohio country, and it's been nonstop development ever since.

Once, unchecked exploitation of our natural landscape was understandable, if not excusable. To the first pioneers, Ohio must have seemed like a land of inexhaustible bounty. But it isn't, and that fact has been indisputably driven home.

The Buckeye State has lost over 90 percent of its original wetlands—second only to California amongst the fifty states. Wetlands (such as bogs, fens, and wet prairie) that are now great rarities once blanketed large expanses. Much of what is now Buckeye Lake in Licking County was a massive cranberry bog. The 427 acres of Cedar Bog are all that remain of an approximately seven-thousand-acre cedar fen. Grand Lake St. Marys in Auglaize and Mercer counties, the largest man-made lake in the world when it was built, flooded what must have been a spectacular wet prairie. The Great Black Swamp of northwest Ohio has been drained and cropped. Attendant with these losses has been a great diminishment of wetland flora and fauna.

Unimaginable schools of fishes choked pristine Ohio streams in John James Audubon's time. Witness this 1848 account by Samuel

Prescott Hildreth, one of the first settlers to describe the natural history of the Ohio River Valley: "[Streams] were filled with delicious fish in such abundance, that at certain times of the year, the smaller tributaries might be said to have been alive with them." Of the 166 species of fishes documented in Ohio, seven species have vanished. Once, our lands were cloaked in dense old-growth timber that covered 95 percent of the state. Today, only a few small bits of ancient forest remain. Elk, American bison, cougar, gray wolf, and American marten all roamed our primeval landscape.

Flocks of wild parrots once plundered fruit from Ohio trees. Carolina parakeets, gorgeously colored in green, yellow, and red, were the only northern parrot. The last were seen in Ohio in 1862, and they went extinct by the early 1900s. Forests of virgin timber teemed with indescribably huge flocks of passenger pigeons; populations numbered in the billions. No one who witnessed these hordes, which might take days to pass a single point and break limbs off trees with the weight of roosting birds, would have predicted they would be extinct by 1914.

What wasn't forest was mostly prairie—about fifteen hundred square miles of it. Stands of big bluestem and Indian grass stretched to the horizon, creating grassy seas that were a daunting sight to westward traveling settlers. In 1837 a man named Deere released a new invention, the chisel plow. Our oceans of prairie grasses and wildflowers were quickly converted to the modern prairie—corn, soybeans, and wheat. Today, less than 1 percent of original prairie survives.

Of the more than eighteen hundred species of native plants thus far documented in Ohio, 601 are listed as endangered, threatened, or potentially threatened—one-third of our flora. Statistics are similar for most animals. Our freshwater mussels have been especially hard-hit. Degradation of water quality has led to the extirpation of nearly 20 percent of our original species. Of the remainder, over half are now state-listed. No one has a handle on many insect families and what losses may have occurred in those groups. There just hasn't been time to study them thoroughly.

Loss of even a single organism can have impacts that ripple through entire ecosystems. Sometimes the effects are immediate and obvious; other times a species's disappearance may not be noticed. In the words of conservationist Aldo Leopold, "To keep every cog and wheel is the first precaution of intelligent tinkering."

Ohio's Biodiversity

Ohio is a medium-sized state—the thirty-fifth largest of the fifty—covering about forty-four thousand square miles. A wealth of biodiversity fills our landscape with influences from western prairies, northern bogs, Appalachian mountains, and the mighty Mississippi River via its major tributary, the Ohio River.

Listed below are species totals for most major groups of Ohio organisms:

- Moths—2,500 (estimated)
- Vascular Plants—Approximately 1,800 native species
- Spiders—580+
- Birds—418
- Mosses—411
- Lichens—223 (not including 204 crustose lichens, which are not well known, and many remain to be found)
- Dragonflies—164
- Butterflies—135
- Fishes—134
- Freshwater Mussels—98
- Mammals—51
- Amphibians—34
- Reptiles—32
- Other Insects (countless thousands; no one knows)

Poisonous Flora and Fauna: A common misconception is that poison oak occurs in Ohio. It does not; that species is southern and comes no closer than northern Virginia. Poison sumac is in Ohio, but is rare and local, confined largely to bogs and fens. Poison ivy is abundant and found in every county.

There are three species of venomous snakes in Ohio: copperhead, massasauga rattlesnake, and timber rattlesnake. Only the copperhead is fairly wide-ranging and somewhat common; it is found throughout the unglaciated Appalachian Plateaus and sparingly elsewhere in southern Ohio. The other two snakes are quite rare and local and discussed later in this book. Cottonmouths (sometimes called water moccasins) are often mistakenly attributed to Ohio. This southern species ranges no closer than southwestern Kentucky, and all Ohio reports are misidentifications.

Two well-known spider groups are found in Ohio that can inflict unpleasant bites: recluses and widows. Two species of recluses are known: brown and Mediterranean. Neither is thought to be native nor likely to be encountered. Typically they are found in buildings and probably can't survive Ohio's winters outside. Black and northern widows are found sporadically statewide but are generally rare. Females bite, and the effects can be nasty. Widows normally live in wood piles, rocky debris, under logs, and similar haunts. Exercising common sense, such as not reaching into crannies sight unseen, will probably eliminate most risk of spider bites.

Physiographic Regions and Original Vegetation of Ohio

Sites in this book are divided by the five physiographic regions that occur in Ohio. Each has a unique character in regard to geology and plant communities. The map is courtesy of the Ohio Division of Geological Survey.

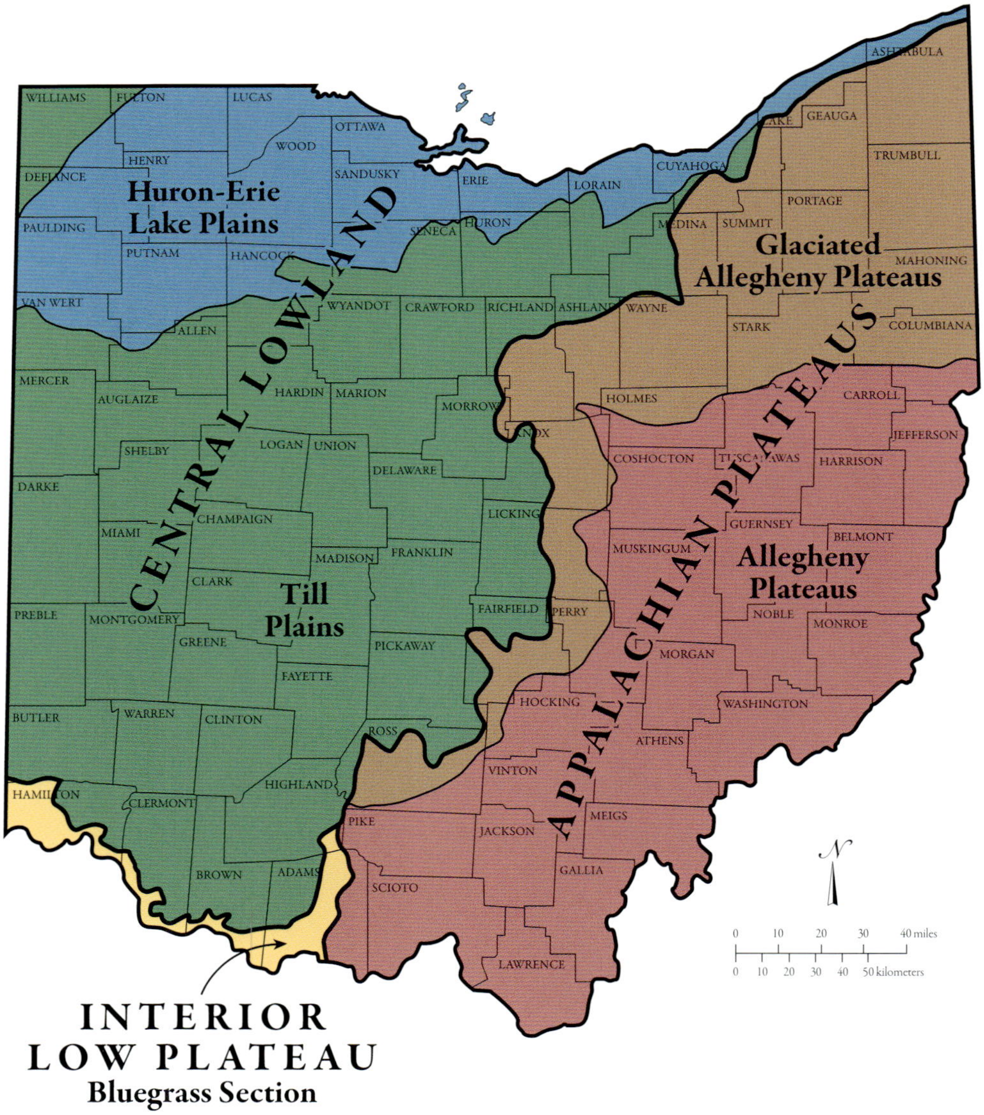

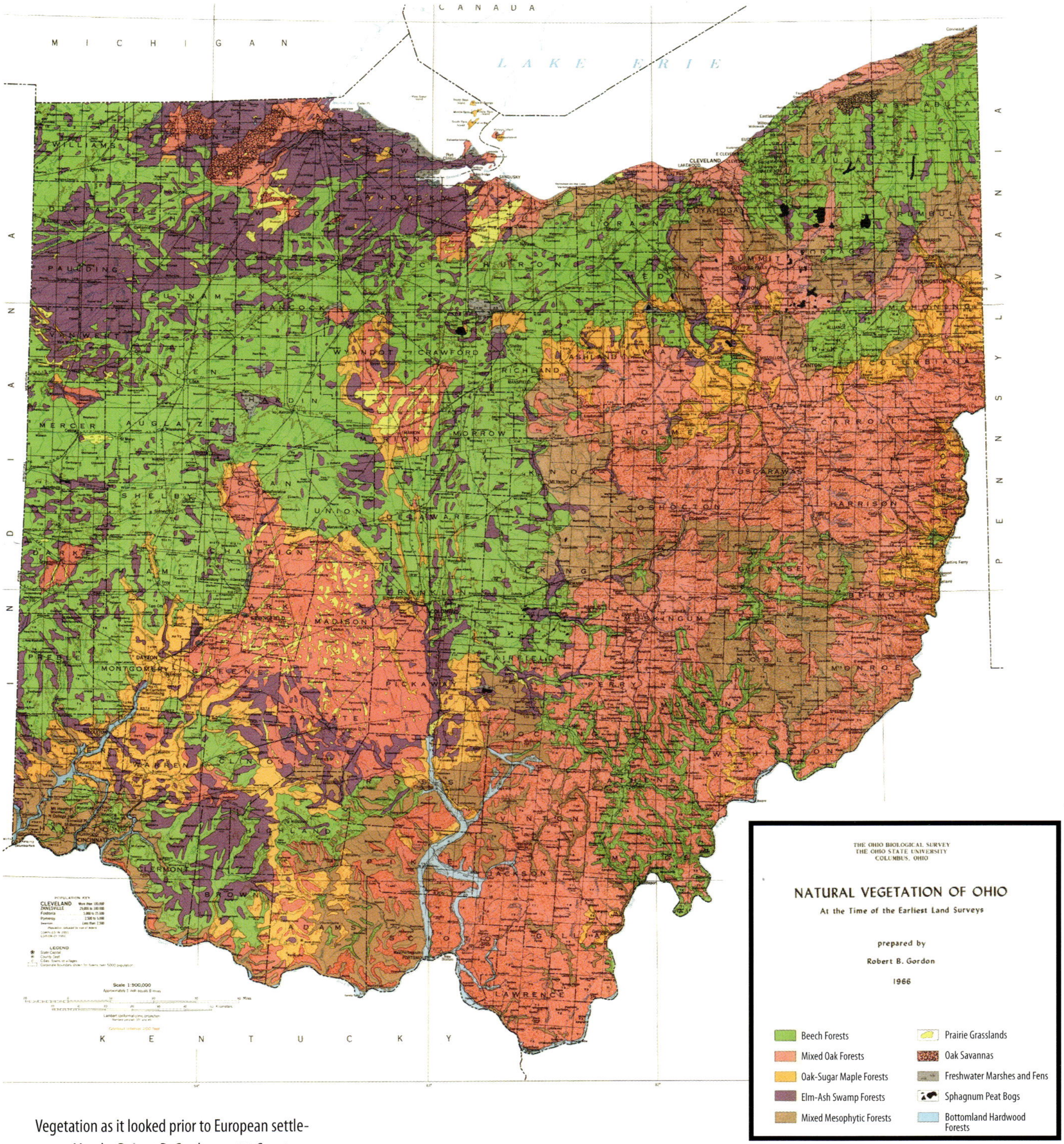

Vegetation as it looked prior to European settlement. Map by Robert B. Gordon, 1966. Courtesy of the Ohio Biological Survey.

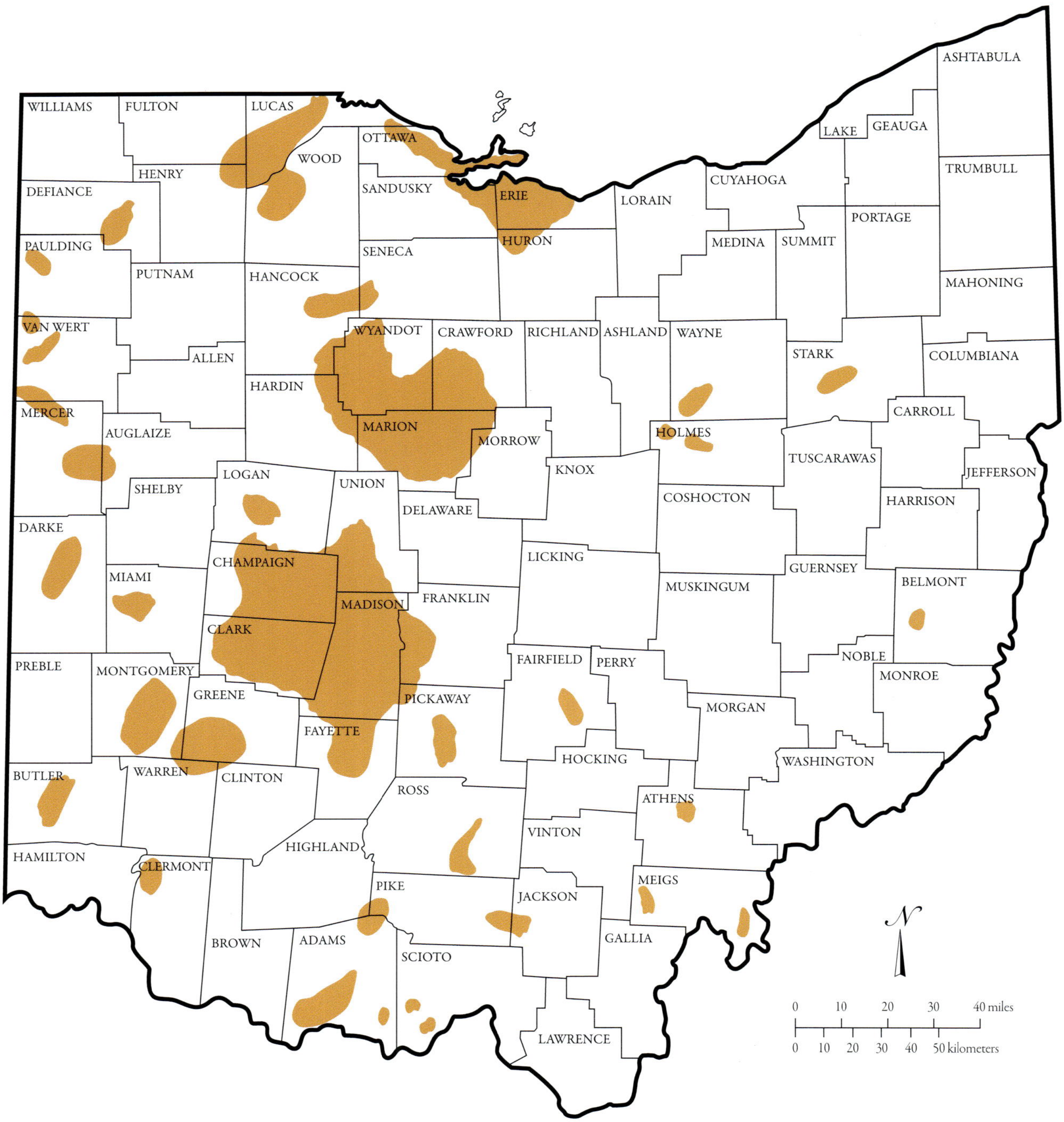

Approximate extent of Ohio's original prairie prior to European settlement. Map created courtesy of Lisa Van Doren, Ohio Division of Geological Survey.

Ohio's Major Habitats

Forests: The largest contiguous forested tracts are in southeastern Ohio hill country. Dry ridges feature oak and hickory, while stream bottoms and lower slopes support mesophytic species such as tulip tree and green ash. The most extensive forests remain in this region, as the hilly topography is not generally conducive to farming or other large-scale development.

Glaciated regions of the state are forested with oak-hickory on well-drained sites, ranging to elm-ash swamps on poorly drained lands. Most forests in glaciated Ohio have been cut and converted to croplands or have been urbanized. Hardly any woods of substance remain in some western counties where agriculture dominates. The once vast Great Black Swamp that covered the northwestern corner of the state has been completely drained and cut over.

Prairie: Approximately four to eight thousand years ago, Ohio's climate was much hotter and drier—an age known as the Xerothermic period. These conditions discouraged trees and encouraged eastward expansion of prairies from the Great Plains. At the time of European settlement, an estimated fifteen hundred square miles of prairie was present. Scattered pockets still occur well to the east of traditionally delineated prairie regions, such as in Athens and Holmes counties.

Most prairies occurred in large, contiguous regions such as the Darby Plains west of Columbus, the Pickaway Plains south of Circleville, and the Sandusky Plains near Marion. After machinery was invented that could cut thick prairie sod, this habitat soon vanished as settlers realized that lush crops could be grown in the rich soil. Today, less than 1 percent of our original prairie remains. The prairies of the Bluegrass Region, which encompass parts of Adams, Highland, and Pike counties, predate the Wisconsin glacier and are much older than other Ohio prairies. These small prairie openings harbor a number of globally rare plants.

Wetlands: At the time of settlement, vast expanses of wetlands covered Ohio. Now Ohio has the dubious distinction of being second only to California in wetland acreage lost—over 90 percent. The most common type was and is swamps, which are wooded wetlands. The largest tract of Ohio swampland was the Great Black Swamp of northwestern Ohio, but much of the glaciated Till Plains was also swamp forest.

Marshes are wetlands dominated by herbaceous plants such as bulrushes and cattails, and the biggest remaining marshlands are the wetlands buffering western Lake Erie. Our second most frequent wetland type, marshes are still commonly found throughout Ohio.

Much of our former prairies were also wetlands, vegetated by specialized wetland prairie plants such as wheat sedge and prairie cord grass. Wet prairie verged into fens in the wettest, most alkaline areas. Fens are "sweet bogs," wetlands fed by cool artesian groundwater and populated by distinct plant life restricted to fen habitats.

Bogs are glacial kettle lakes filled with acid-loving plants growing on a substrate of *Sphagnum* moss. This wetland type was never plentiful in Ohio, at least in the last several thousand years, and is mostly found in the northeastern corner of the state.

Unique among Ohio wetlands is Cedar Bog (pages 39–41), which is a white cedar–dominated wetland. This is a habitat normally found far to the north, and which probably never occurred elsewhere in Ohio, although Cedar Bog was far larger at one time.

Lake Erie: The second smallest of the Great Lakes buffers most of Ohio's northern border and has a profound influence on our biodiversity. Lake Erie ecosystems support some of the most important migratory bird habitat in the Midwest and it is the most biologically rich of the five Great Lakes. Open lake waters attract masses of water birds and support world-class fisheries. Shoreline habitats are varied, from sand beaches to cliffs to marshes.

Oak Openings: The old beach ridges and dunes of preglacial Lake Warren, Lake Erie's predecessor, cover about 130 square miles west of Toledo. Now found mostly in Lucas County, the Oak Openings originally covered three hundred square miles. Ohio's greatest concentration of rare species occurs here.

Rivers: Ohio has over sixty thousand miles of streams, and they play a vital role in supporting our diverse flora and fauna. Our largest river is the Ohio, of which about 451 miles serve as the state's southern border. Many species of animals and plants migrated northward into Ohio via this Mississippi River tributary, and a number of them occur no farther north. Large lock and dam projects designed to facilitate shipping have largely destroyed the natural character of the Ohio River.

The Teays River was a prehistoric drainage that flowed northward into Ohio from southern Appalachian regions up until about two million years ago. Over the Teays' existence, numerous Appalachian plant species such as rhododendrons and magnolias migrated north along the river and can be found in southern Ohio to this day.

The Lake Erie drainage St. Marys River of western Ohio forms a natural connection with the Mississippi drainage Wabash River, and serves as a vital conduit for animals and plants to migrate between the drainages. Many species entered the Great Lakes drainage through this natural portal. Big Lake Erie tributaries such as the Maumee, Sandusky, and Grand rivers provide critical spawning habitat for lake fishes such as walleye.

Ohio founded the nation's first scenic rivers program in 1968 and has long been at the forefront of river conservation. To date, 754 river miles on parts of twenty-three streams have been designated as scenic rivers. This program is administered by the Ohio Division of Natural Areas and Preserves.

Selection of Sites and Other Notes

Few seasoned Ohio naturalists would dispute that the places included in this book belong in a listing of our best remaining natural areas. There are plenty of other worthy candidates, though. Space precludes detailing them all. We wanted to feature areas that exhibit the best attributes of the habitat or ecosystem in question. The one exception is Tri-Valley Wildlife Area, which is the product of strip mine reclamation and is now a mammoth artificial grassland. After debate we decided that its importance to grassland birds warrants its inclusion. Tri-Valley also represents a habitat of which as many as 250,000 acres are now found in southeastern Ohio, with important ramifications for bird conservation.

All of the areas in this book are available for public access, although permits or special arrangements may have to be made for some. There are many other outstanding natural areas that are privately held, and we didn't want to get into the complexities of publicizing such sites.

Fifteen agencies and organizations own the sites discussed in this book. Collectively, they protect examples of every major habitat type in Ohio and nearly 100 percent of our flora and fauna.

In the main text, only common or English names of plants and animals are used. Fortunately, animals have increasingly had their English names standardized, and there is less ambiguity when referring to them without using a scientific name. Not so with plants, and botanists will often be puzzled if no scientific name is provided, particularly for groups like grasses and sedges. Because of the hundreds of species of flora and fauna mentioned in the text, we wanted to avoid cluttering it by pairing scientific names with every organism mentioned. A separate index lists all animals and plants alphabetically by English name, along with their scientific name.

Finally, we encourage you to visit these places and view the greatest natural resources that Ohio has to offer.

Acknowledgments

A book like this is the result of many years of fieldwork, and learning from others. Scores of naturalists and biologists have shared not only their knowledge with the author and photographer but also hidden caches of natural lore. We are grateful for their help, and hope that no omissions have been made.

Especially helpful in directly aiding with this book were: Jim Bissell, Tim Matson, John Pogacnik, Dan Rice, and Jeff Wolfinger. Staff of the Ohio Natural Heritage Program provided much valuable data on Ohio's rare organisms and habitats: Tom Arbour, Barb Burkholder, Rick Gardner, Butch Grieszmer, Erin Hazelton, and Greg Schneider. Invaluable assistance with maps came from the Ohio Division of Geological Survey, particularly Lisa Van Doren for her map-making skills and E. Mac Swinford for graciously making resources available to us. Bruce Motsch of the Ohio Department of Natural Resources, Division of Real Estate and Land Management was very helpful in providing maps as well.

The following individuals have spent time in the field with us or otherwise helped us to better understand Ohio's natural history. Their enthusiasm for nature and passion for conservation contributed to the contents of this book, and their expertise made it a better product: Ian Adams, Mac Albin, Barbara Andreas, Tom Bain, Joe Barker, Tom Bartlett, Chris Bedel, Dana Bollin, Aaron Boone, Dan Boone, Renee Boronka, Bill Bosstic, Kevin Bradbury, Richard Bradley, Doug Brewer, Lisa Brohl, Deb Brill, Jennifer Brumfield, Don Burton, Ted Cavender,

Phil Chaon, Delores Cole, Tom Cooperrider, Mix Cooperrider, Gary Coovert, Holly Coovert, Dave Corbin, Allison Cusick, Jim Davidson, Jim Decker, Guy Denny, Mark Dilley, Dave Dister, Jim Dolan, Judy Dumke, Doug Dunakin, Micki Dunakin, Jon Dunn, Vic Fazio, Mike Flynn, John Freudenstein, Jim Fry, Bob Gable, Brian Gara, Robert Glotzhober, Steve Gray, Gary Haase, Ben Hamilton, Joe Hammond, Rob Harlan, Cheryl Harner, Bob Harter, Pat Hayes, Lori Hayes, Dave Horn, John Howard, Bob Jackse, John Jaeger, Terry Jaworski, Kenn Kaufman, Kim Kaufman, Ned Keller, Marc Kibbey, Charlie King, Peter King, Ethan Kistler, Paul Knoop, Jason Larson, Greg Lipps, Tami Longaberger, John Mack, Bernie Master, Martin McAllister, Bret McCarty, Jim McCarty, Richard McCarty, Martha McCormac, John McCormac, Mike McCormac, Charlotte McCurdy, Jack McDowell, Debbie McFadden, Gary McFadden, Steve McKee, Joe McMahon, Karen Menard, Jane Meszaros, Eric Metzler, Len Mikles, Greg Miller, Lucy Miller, Dave Minney, Debora Moon, Marshal Moser, Rick Nirschl, Dave Nolin, John O'Meara, Marilyn Ortt, Al Parker, Brian Parsons, Bruce Peterjohn, Ralph Pfingsten, Bob Placier, Carolyn Platt, Steve Pollick, Ralph Ramey, Anton Reznicek, Jenny Richards, Stan Richards, Larry Richardson, Emliss Ricks, Dave Riepenhoff, Larry Rosche, Steve Roloson, Dan Sanders, Robert Sanford, Jen Sauter, Marty Sedluk, Judy Semroc, Ray Showman, Steve Smith, Su Snyder, Tim Snyder, Al Staffan, David Strawberry, Stanley Stine, Warren Stoutamire, Ed Stroh, Ronald Stuckey, Dave Swanson, John Switzer, Merrill Tawse, Bill Thompson III, Elliot Tramer, Roger Troutman, Warren Uxley, Michael Vincent, Edward Voss, Joel Wachtel, Tim Walters, Ben Warner, John Watts, Tom Watters, Art Weber, Patricia West, Pete Whan, Kelly Williams-Sieg, Steve Willson, Doug Wynn, Julie Zickefoose, Phil Zito, and Mark Zloba.

Special thanks to all Ohioans who enjoy the outdoors and actively support conservation.

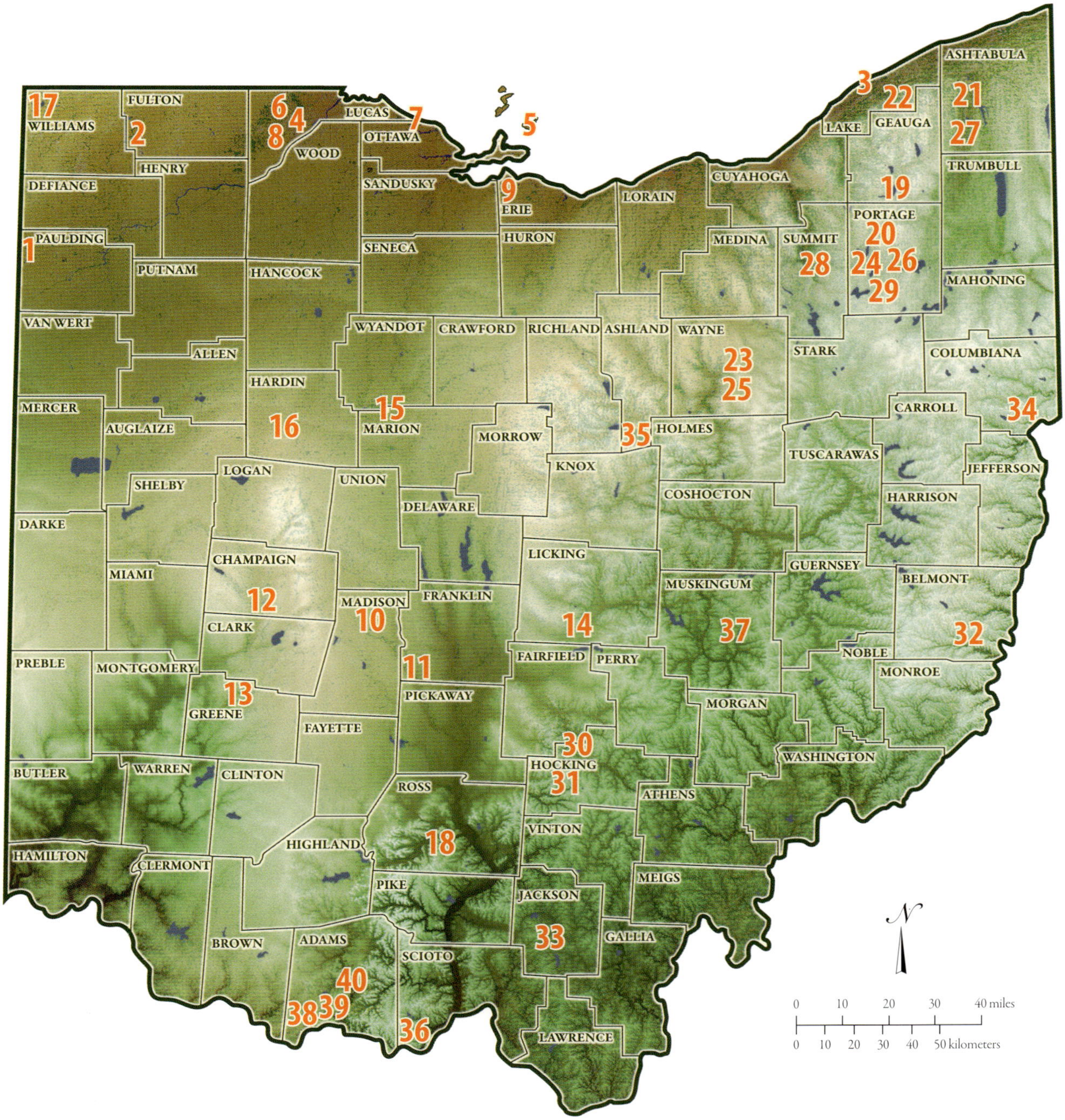

For a complete listing of regions for each corresponding number, see specific maps at the beginning of each section.

Appalachian Plateaus

Clear Creek Metropark (30) • Conkles Hollow State Nature Preserve (31) •
Dysart Woods (32) • Lake Katharine State Nature Preserve (33) • Little Beaver Creek (34) •
Mohican State Forest (35) • Shawnee State Forest (36) • Tri-Valley Wildlife Area (37)

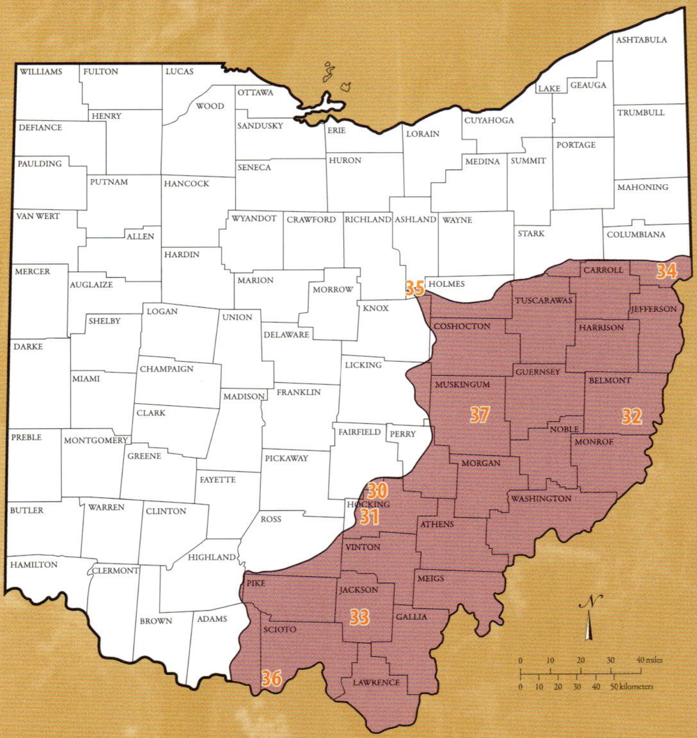

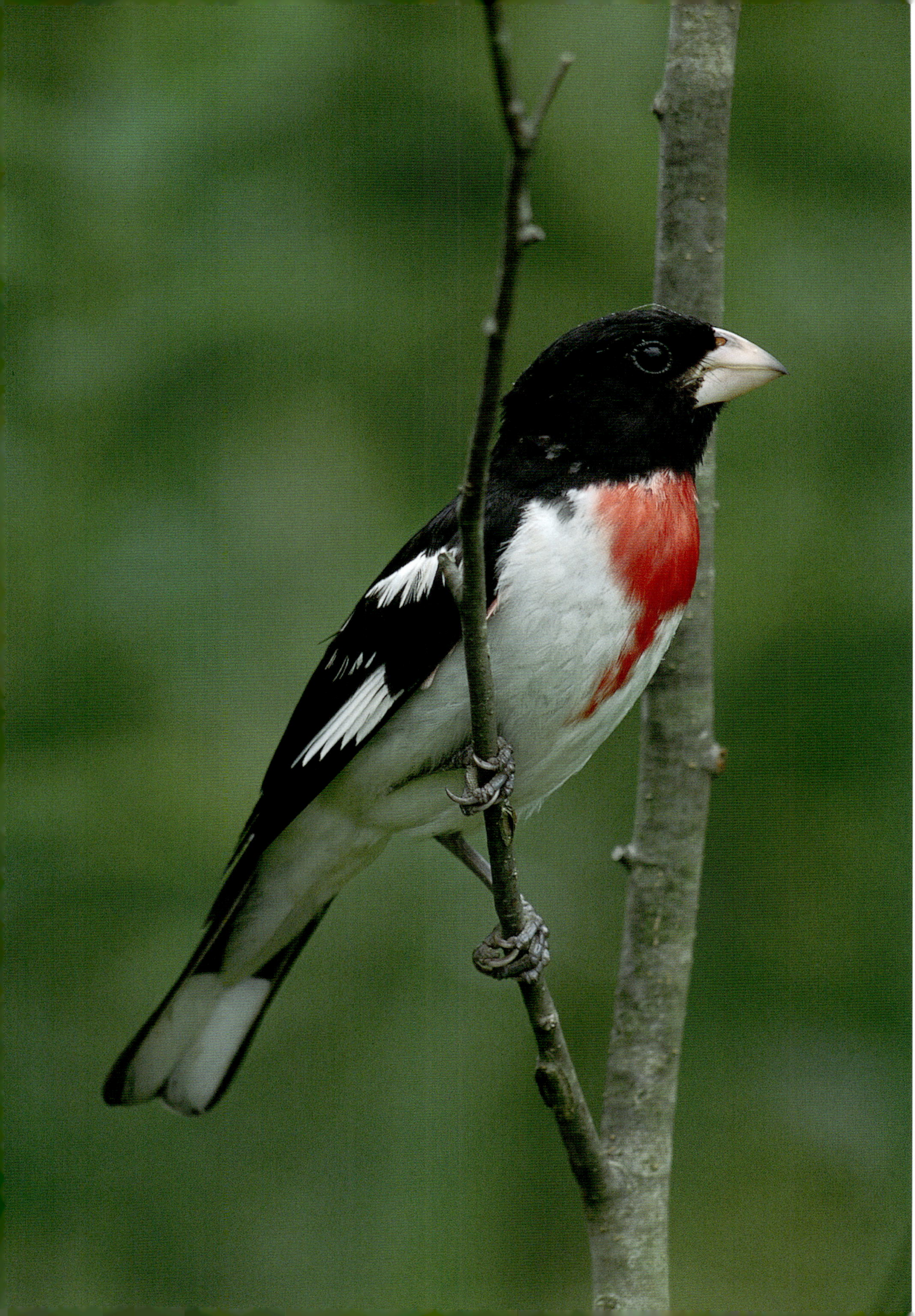

Plant communities from every compass point converge at Clear Creek. This impressive valley was scoured through highly erosive Mississippian-age sandstone over millennia by nature's great mason, water. Located on the northern fringe of the Hocking Hills and only forty minutes from Columbus, Clear Creek is one of the most beautiful and diverse ecosystems in Ohio.

Northern influences are obvious. A boreal conifer, Eastern hemlock, blankets the north-facing slopes of Clear Creek, their boughs casting deep shade over the forest floor. Interspersed is yellow birch, a tree of Appalachian coves and northern woods, with its flaky bark and twigs that taste of wintergreen. Disjunct relicts, a number of species of boreal birds rarely found this far south nest in the valley. Beautiful flutelike melodies of hermit thrushes join the songs of other northerners, such as blue-headed vireo and Canada warbler. A stone's toss away, the south-facing stream slopes provide a world of contrast to dense hemlock groves. Deciduous forests dominate on these sunnier slopes, characterized by trees like white and red oak, shagbark hickory, tulip tree, and sourwood. The breeding birds differ radically, too. Southerners like worm-eating and Kentucky warblers, and occasionally summer tanager, are found. Black vultures, near their northern limits at Clear Creek, are common—one of few Ohio strongholds.

Appalachian flora, at its western limits in the valley, is conspicuous. Thickets of mountain laurel crest rocky summits, and a few patches of rare great rhododendron grow on steep slopes. Sourwood and

Clear Creek Metropark

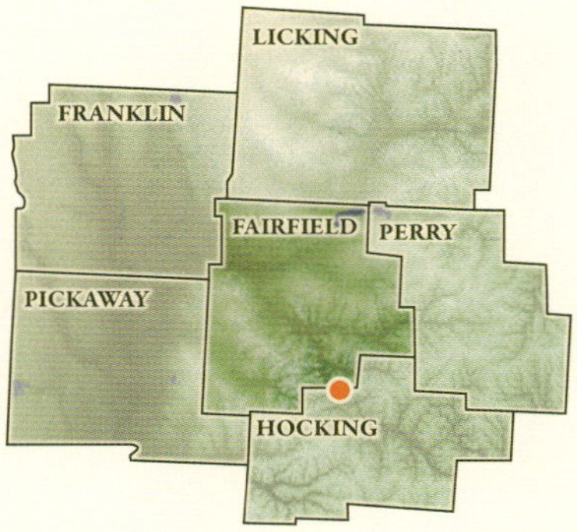

ACREAGE: 5,252

OWNERSHIP: Columbus and Franklin County Metroparks

ACCESS: Open during posted hours; good network of trails

WEB SITE RESOURCES:
http://www.metroparks.net/

http://www.ohiodnr.com/dnap/location/clearcreek.htm

http://www.ohiobirds.org/birdingsites/showsite.php?Site_ID=46

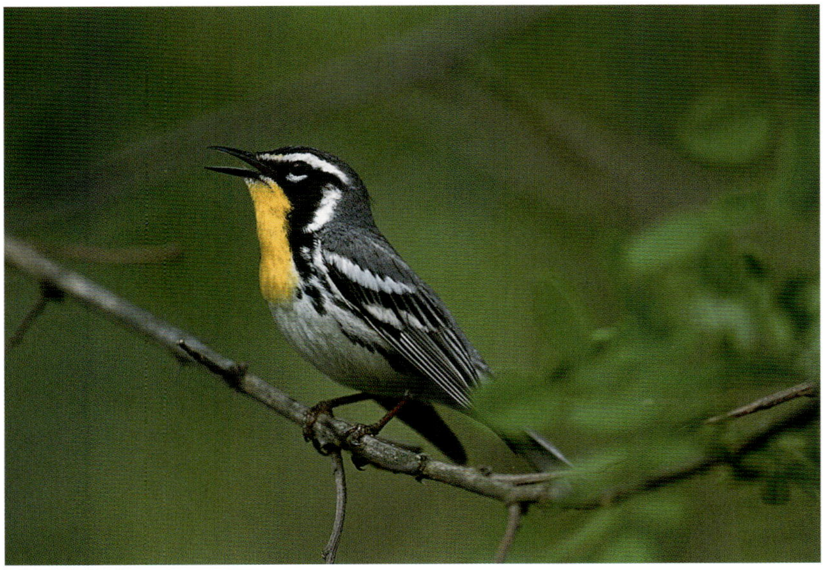

Most yellow-throated warblers in Ohio are associated with sycamore trees; indeed, the bird was once known as sycamore warbler. One of our first neotropical species to return, its clear cascading notes can be heard by late March.

Opposite: Beautiful rose-breasted grosbeaks are common breeding species of riparian corridors. Nesting in the midcanopy, their rich robinlike song can be heard up and down the creek.

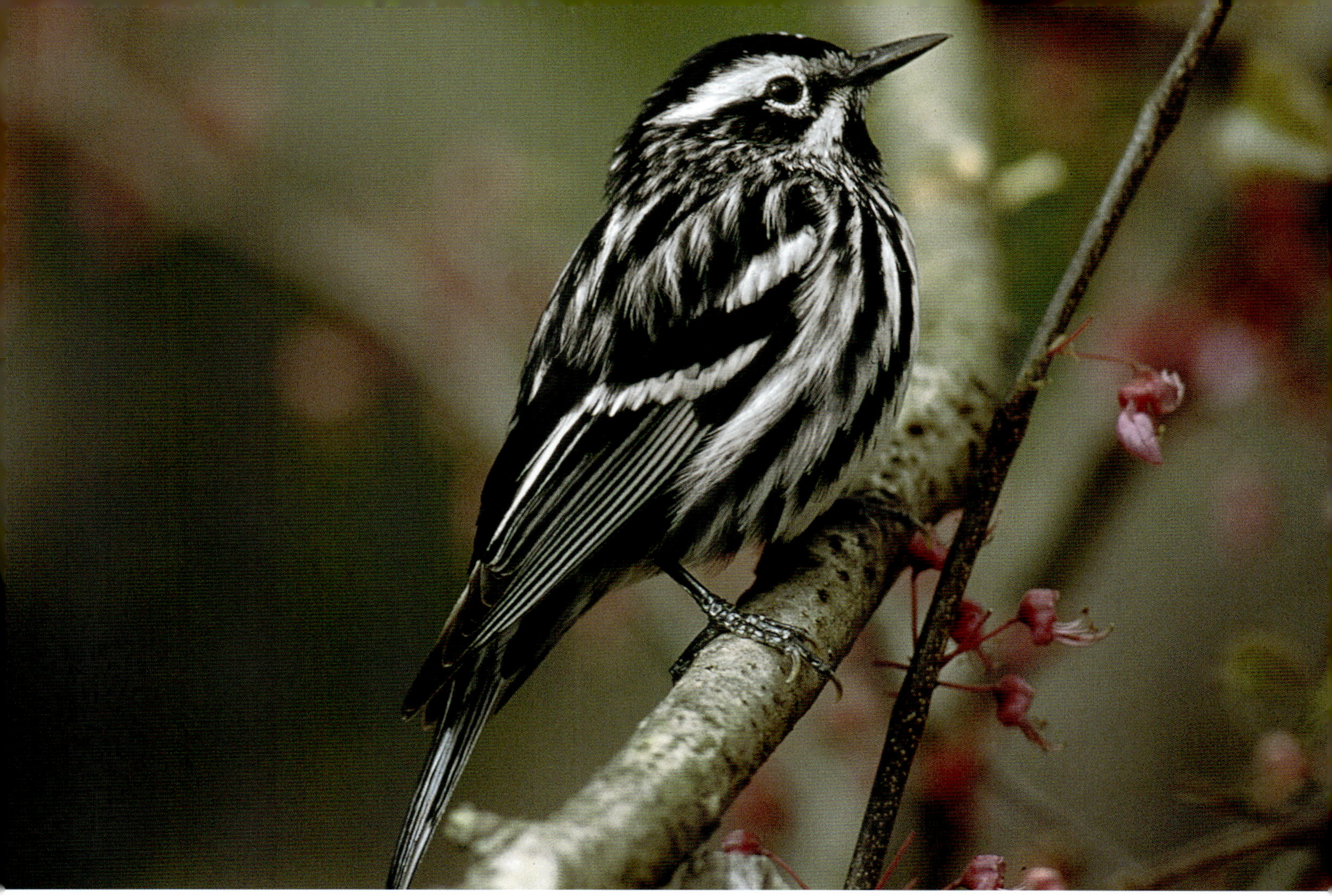

Black-and-white warblers forage along the trunks and limbs of trees in the manner of a nuthatch or brown creeper, aided by a hind claw much longer than that of other warblers. Interestingly for such an arboreal species, they place their nests on the ground—one of nine of the twenty-five regularly nesting Ohio warblers to do so.

chestnut oak, Appalachian trees found no farther west in this region, are common on xeric ridges. A treat is finding patches of pink lady's-slipper in the upland woods.

Streamside habitats feature ghostly white sycamore trees, often underlain by thick carpets of scouring-rush, an odd reedlike fern and one of our most primitive plants. Its relative the ostrich fern, almost unknown in central Ohio, also grows sparingly on stream terraces.

Sandstone outcrops abound, and these rocky promontories support interesting plants like wild columbine, rock polypody, and early saxifrage. A can't-miss rock is Leaning Lena, a sandstone slump block that tumbled from the cliffs above long ago. Clear Creek Road passes underneath the wildly tilted cabin-sized monolith, inspiring passersby to hope it doesn't choose that moment to complete its fall. Similar sandstone blocks litter cliff bases throughout the valley; the results of water erosion of softer underlying rocks.

Clear Creek Valley has been studied by naturalists for many decades. Legendary biologist Edward S. Thomas had a cabin in a side valley he dubbed Neotoma, named for the genus of the Allegheny wood rat. That rare mammal no longer occurs, but plenty of other biodiversity does. To date, about eight hundred species of plants have been documented in Clear Creek's varied habitats. Over 110 species of birds nest in and around the valley, including several of Ohio's rarest breeding species.

First-time visitors would do well to hike the Hemlock Trail, which offers a taste of many of the valley's habitats. A trip to Clear Creek won't disappoint, at any season.

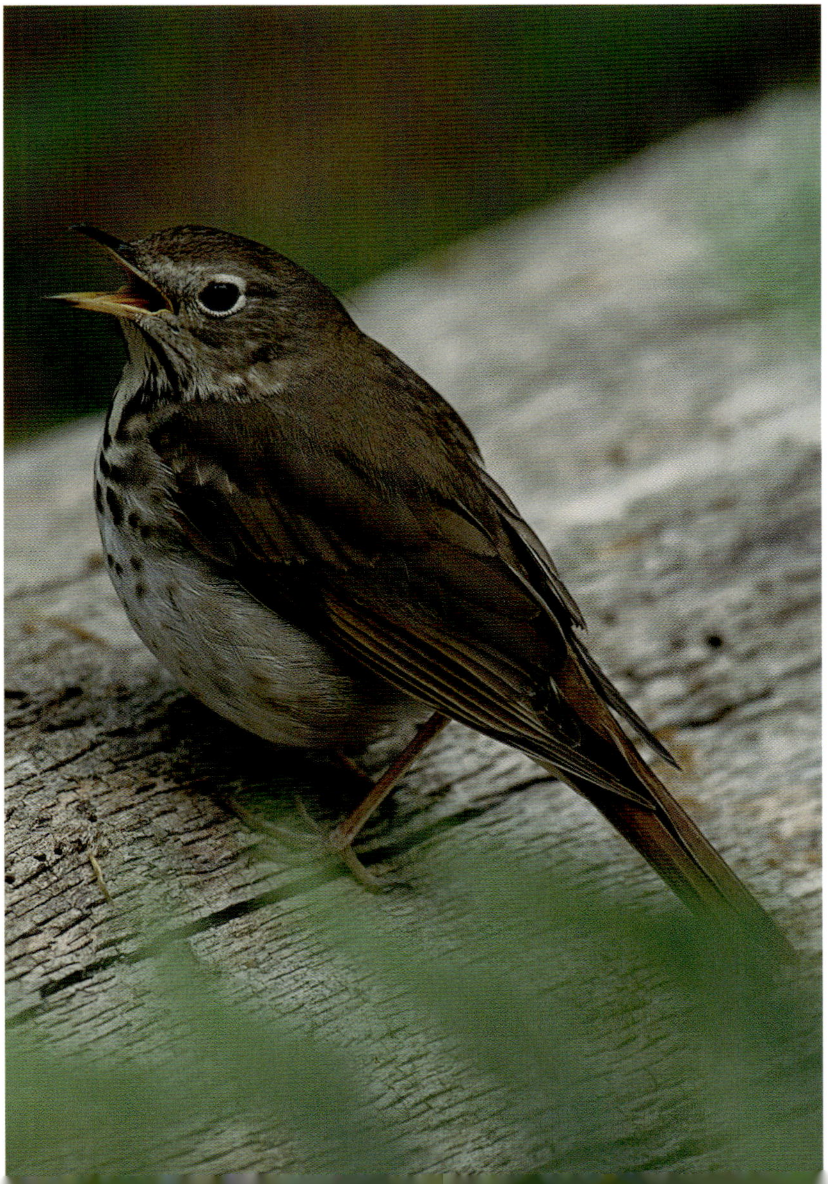

The ethereal song of the hermit thrush is an unforgettable melody. Along with a small suite of species that includes Canada warbler and winter wren, the thrush is a boreal disjunct.

Conkles Hollow State Nature Preserve

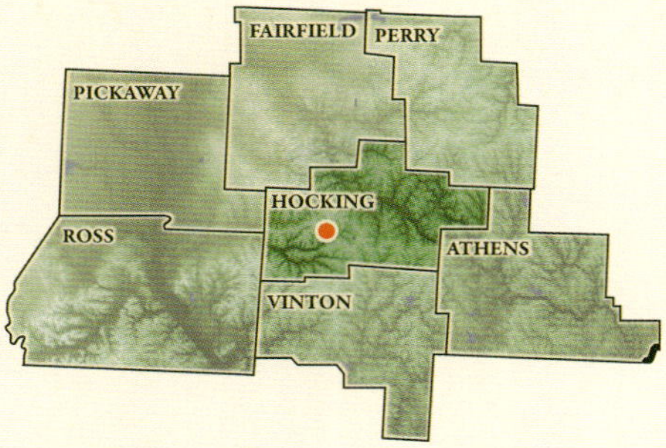

ACREAGE: 87 acres

OWNERSHIP: Ohio Division of Natural Areas and Preserves

ACCESS: Open during posted hours

WEB SITE RESOURCES:
http://www.dnr.state.oh.us/dnap/location/conkles_hollow.html

http://www.ohiobirds.org/birding-sites/showsite.php?Site_ID=62

http://www.1800hocking.com/

Opposite: One of Ohio's most inspiring vistas is the view from the upper-rim trail overlooking Conkles Hollow. The tallest cliff in the state, a vertical sandstone face extends nearly two hundred feet above the valley. In autumn, a tapestry of oak, hickory, sourwood, maple, tulip tree, and other trees tint the forest in rich hues of red, yellow, and orange.

In Hocking Hills, a region renowned for spectacular scenery, Conkles Hollow stands out. Countless thousands of visitors have hiked the gorge, awed by massive sandstone formations, lush beds of ferns, and thick stands of hemlock, all conspiring to create the illusion of being in Canada's north woods.

The story of the gorge dates back about 360 million years ago, when Mississippian shales that comprise much of the Hocking Hills began to form. Numerous geologic upheavals in the intervening millions of years formed the rugged terrain of this region, and nonstop forces of water and gravity carved impressive valleys in softer, more permeable rock strata.

Plant life in the box canyonlike gorge is totally different than along the summits above. Perpetual shade rules in the valley, cast by towering Eastern hemlock trees. The rich humus of stream terraces is lush with ferns, most commonly the fancy fern. It forms luxuriant clumps in profusion and is one of the gorge's most conspicuous botanical features. Other plants of moist shady habitats are easily found, including miterwort, foamflower, and wild stonecrop, the last our only native *Sedum*.

Sharp ears will soon detect summer bird songs not often heard in Ohio. The boreal nature of the hemlock forest attracts many species normally breeding far to the north, such as hermit thrush, black-throated green, Blackburnian, Canada, and magnolia warblers, and blue-headed vireo.

Venture up steep trails to the summits of Conkles Hollow and a completely different woodland awaits. Dry oak-hickory forests rim the canyon below. Barren rocky precipices along the summits host heaths such as deerberry, huckleberry, and our only tree in the family, sourwood. The other Ohio habitat in which heaths dominate are peat bogs. Dry, acidic ridges of the unglaciated hill country are their ecological counterparts.

A little-studied facet of Ohio's natural history is lichens, which abound at Conkles Hollow. Lichens are the symbiotic pairing of a moss and a fungus, and oak trunks and sandstone rocks provide excellent substrates for their growth. Several dozen species occur in and along the gorge. The rarest may be the distinctive pink-dot lichen, found in only a few other locales. Easily dismissed as crusts on rocks or trees, lichens serve several important ecological roles. They begin the long process of rock erosion by collecting moisture and stimulating stress fractures. Insects such as lacewing larvae hide among them, and gnatcatchers and hummingbirds shingle their nests with larger

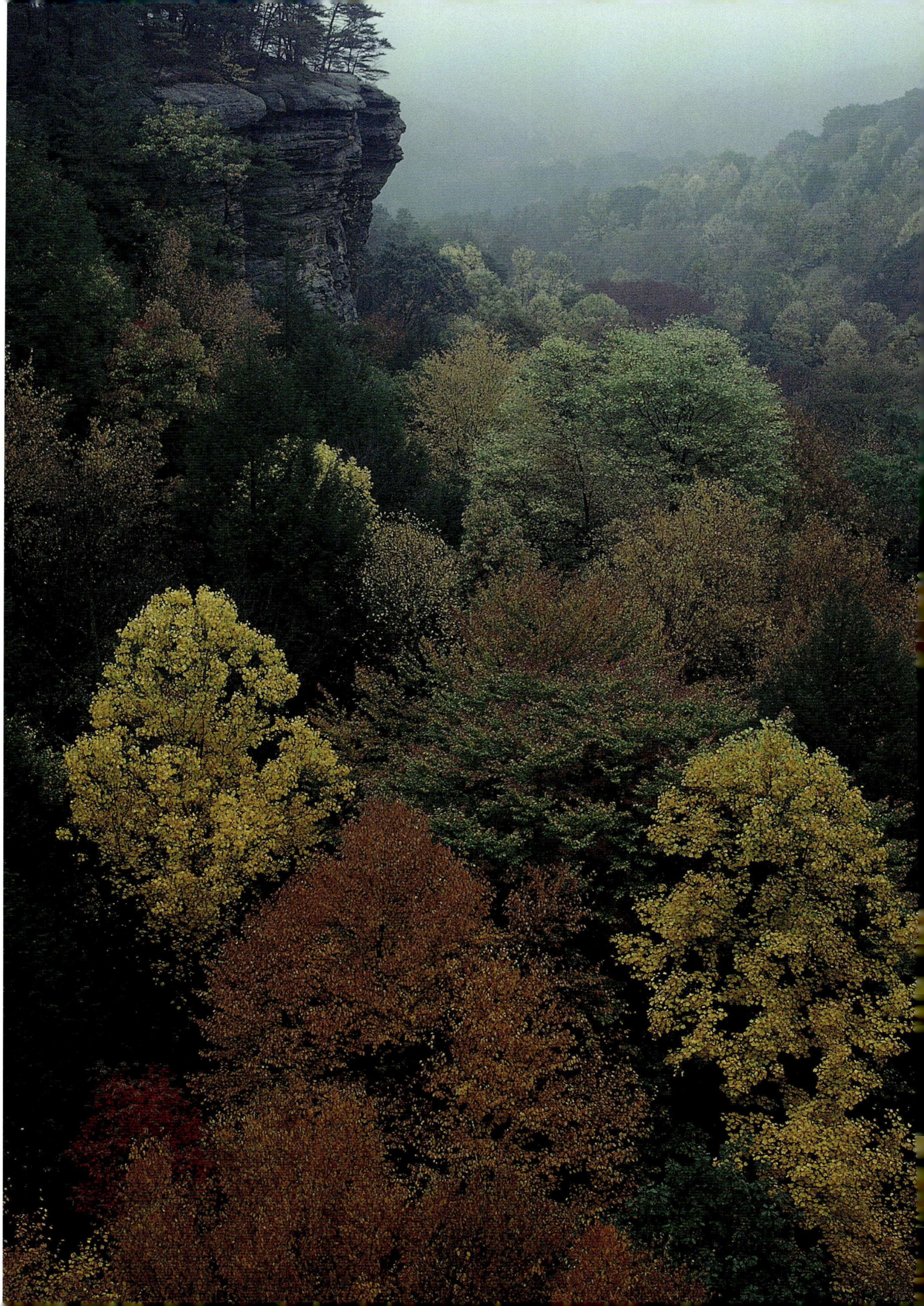

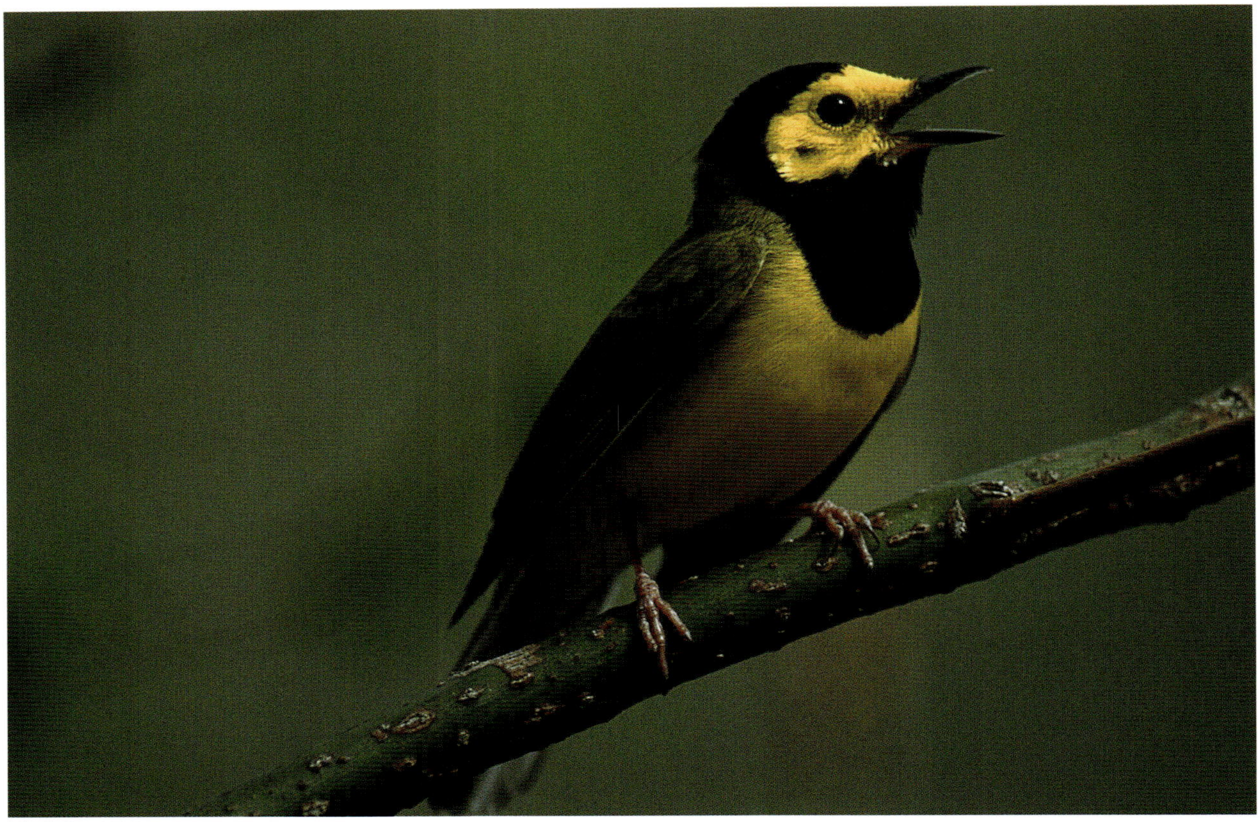

Above: Hooded warblers are denizens of the dense, shady forest understory. As an adaptation to the gloom of their heavily forested haunts, hooded warblers have evolved the largest eyes of any warbler species.

Right: Beautiful Blackburnian warblers are one of Ohio's rarest breeding species. Near the southern edge of their nesting range in Ohio, they occupy only the highest quality hemlock gorges.

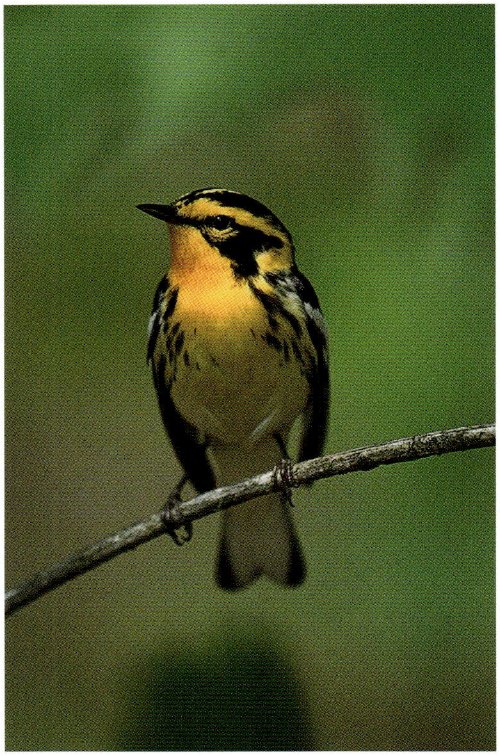

species like greenshield lichen. Lichens also are sensitive air-quality indicators, among the first groups of organisms to vanish if pollution becomes oppressive. American Electric Power, the energy giant, has long used lichens as bio-indicators to gauge plant emissions.

Interesting flora aside, the climb to the top is justified by the sensational views. The tallest cliffs in the state are here, and vistas over extensive surrounding forests are unsurpassed in the region.

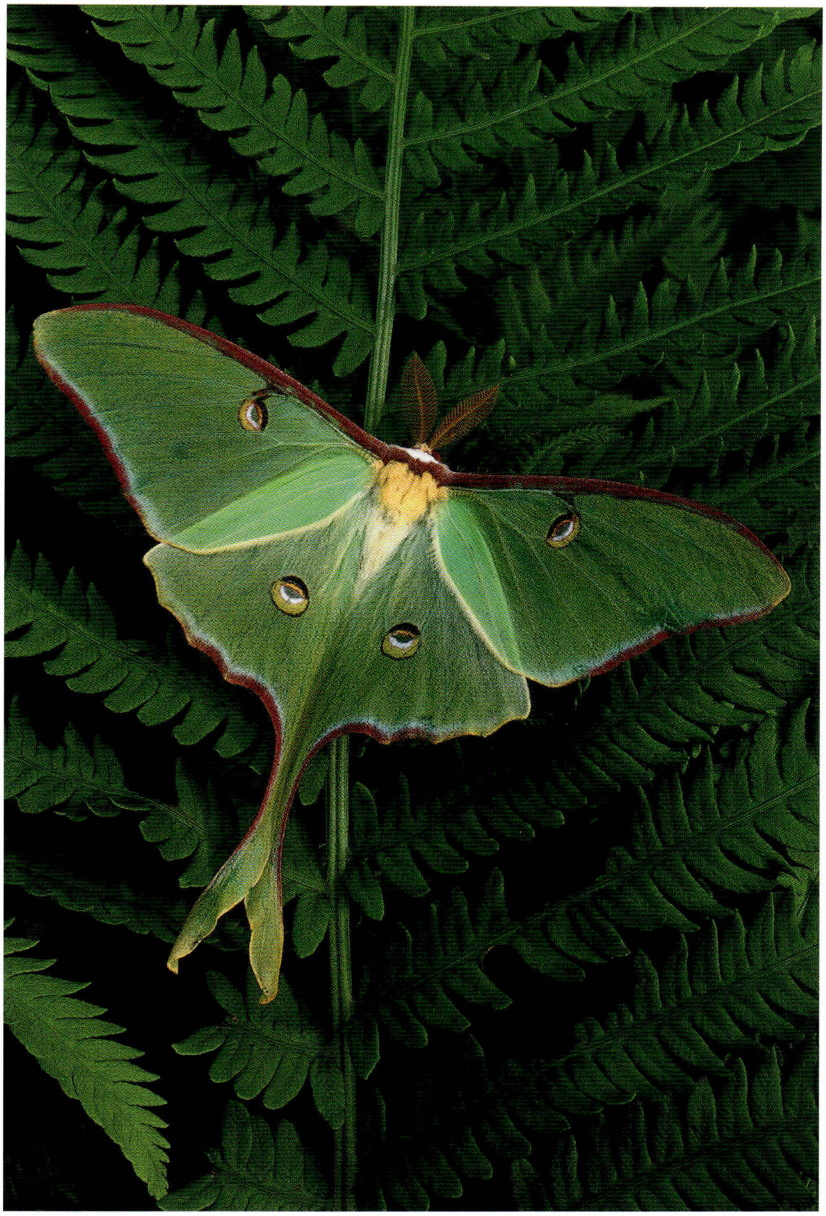

There are about twenty-five hundred species of moths in Ohio. One of our most beautiful species is the luna moth. The caterpillars feed mainly on cherry, hickory, and walnut.

Dysart Woods

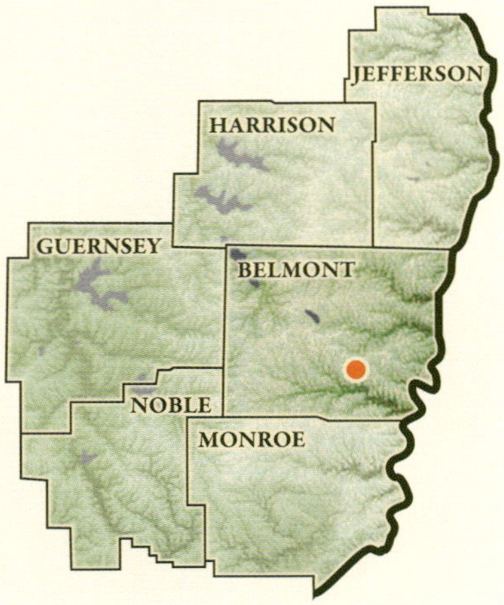

ACREAGE: 50

OWNERSHIP: Ohio University

ACCESS: Open to the public any time

WEB SITE RESOURCES:
http://www.plantbio.ohiou.edu/epb/facility/dysart/dysart.htm

Dysart Woods is a museum piece of giant timber. This magnificent woodlot offers a glimpse back to the presettlement wilds of southeastern Ohio. Some of the largest trees left in the state occur here. Massive red and white oaks and tulip trees push 140 feet into the sky and attain diameters exceeding four feet. This small tract of woodland is perhaps the least disturbed of any forest remaining in the state.

Visitors to Dysart will be instilled with a sense of awe. Trees that weigh many tons and rise to the height of a fourteen-story skyscraper are the eastern counterparts of western redwoods and almost defy belief. It's a testimony to human ingenuity and energy that we managed to clear the landscape of woodland behemoths so rapidly after settlement.

Huge trees create multidimensional habitats on a vertical plane. Their roots play host to specialized fungi and odd hemiparasitic plants like beech-drops and squaw-root, as well as bind the soil and prevent erosion. Massive trunks provide substrates for myriad arboreal lichens, mosses, and fungi, some of which provide habitat for insects, material for bird nests, and, in the case of lichens, excellent barometers of air quality.

The crowns of such mammoths have worlds of their own. Many species of birds, such as cerulean warblers and pileated woodpeckers, require mature timber. Untold varieties of *Lepidoptera* (butterflies and moths) live and forage in the leaves. One rare butterfly, the early hairstreak, may be more common than we think. Apparently it is reluctant to leave sky-high leafy canopies. Massive American beech and

Right: A tree-top dweller, the cerulean warbler sings a buzzy, spirally ascending song. Ohio supports some of the best remaining breeding populations. Cutting of our eastern deciduous forest has eliminated much breeding habitat for this imperiled species.

Opposite: Dysart Woods is renowned for containing some of the largest trees in Ohio. These giant red oaks are several hundred years old. Primeval forests such as Dysart can be counted on one hand. Here, a visitor can get a taste for what Ohio's early pioneers would have seen.

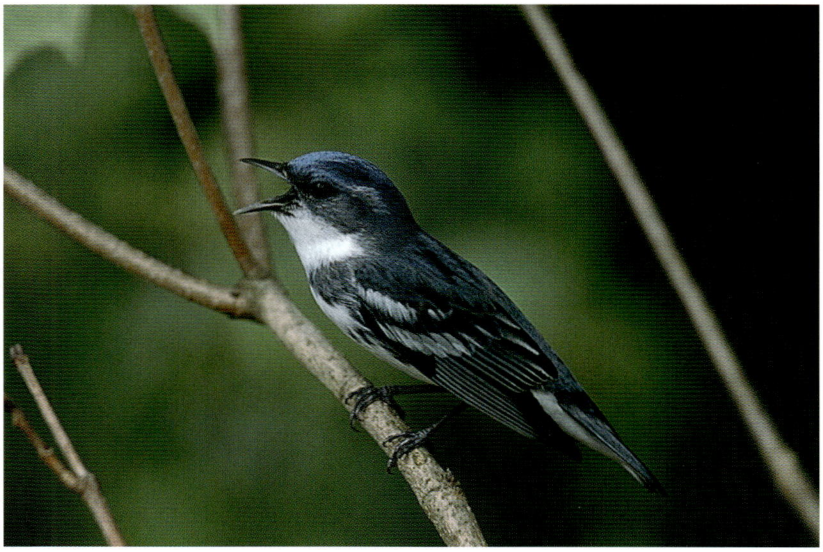

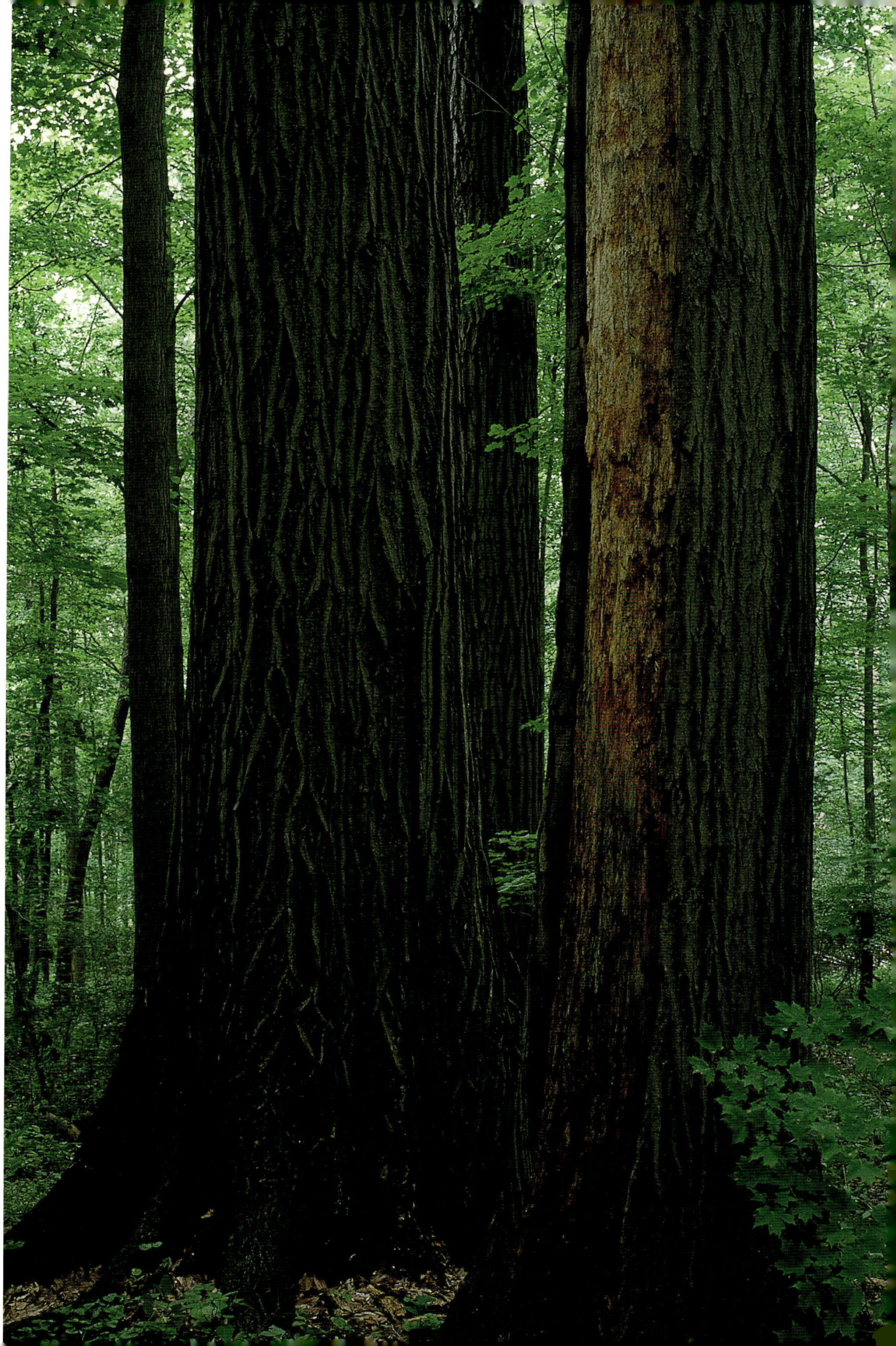

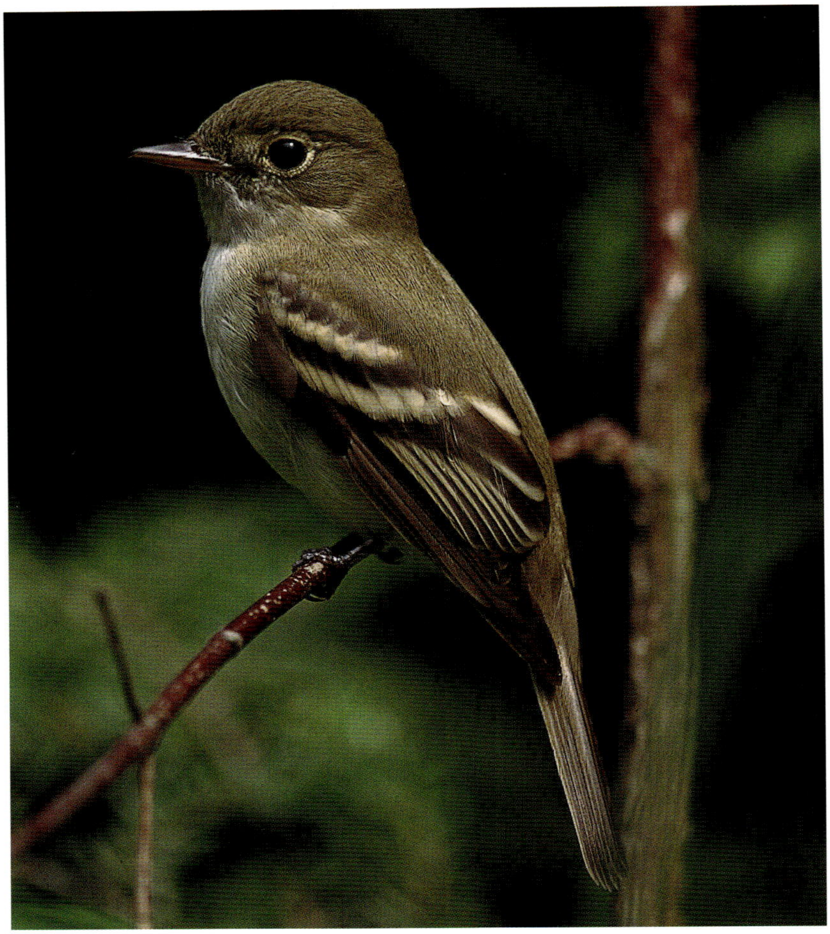

Emitting a loud, explosive song that sounds like *peet-sa,* the Acadian flycatcher is a common mid-canopy breeding bird of rich deciduous eastern forests.

oaks provide bumper crops of mast, which in turn fuel many species of birds and mammals—everything from gray squirrels to wild turkeys.

Sylvan colossi as are found at Dysart once covered over 90 percent of Ohio. Now, only an almost immeasurably small fraction of a percent of old-growth woodland remains. One would think that we would protect these remnants from any intrusion, without debate or caveat. Unfortunately, as is often the case in southeastern Ohio, property owners of surface lands don't own subsurface mineral rights. The owners of the latter are free to access their holdings, which in the case of Dysart Woods involve subterranean coal seams.

After a prolonged legal battle, environmentalists lost a fight to prevent a coal company from boring under Dysart to extract the combustible mineral. Claims have been made that a technique known as longwall mining is noninvasive and won't harm the forest above. Hopefully that's true, but only time will tell.

Knobs and outcrops of Sharon conglomerate sandstone form the substrate at Lake Katharine and make for interesting rock-watching. Trailside cliffs offer educational portals into Ohio's geological past. Sandstone walls often exhibit honeycombed erosion patterns, sedimentary stratification, and recessed rock shelters. Sharon conglomerate formed during the Pennsylvanian subperiod of the Paleozoic era, some three hundred million years past. At that time, vast swampy forests covered the land, and evidence of this past is obvious in the Sharon conglomerate. Various fossilized clams, snails, pebbles, and shells embedded in the rock provide proof that this high and dry land was once wetland.

An excellent trail system offers access to the varied habitats of Lake Katharine. Plant communities change markedly from bone-dry ridgetops to moist creek bottoms. Atop sandstone crests are three of Ohio's four native pines: pitch, shortleaf, and Virginia. Large conifers support breeding pine warblers, a species that must have mature pine stands. The arid, acidic soils of these sandstone ridges are similar to peat bogs floristically, if not in appearance. Heaths dominate, among them several species of low-growing blueberry, wintergreen, trailing arbutus, and sourwood.

Barren cliff faces provide fascinating botanizing. Many species of ferns colonize fissures and ledges of rock formations, including diminutive lobed spleenwort and mountain spleenwort. Sometimes a

Lake Katharine State Nature Preserve

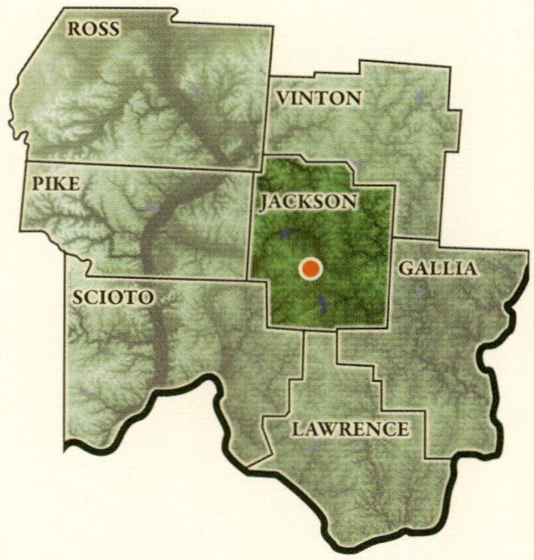

ACREAGE: 2,000

OWNERSHIP: Ohio Division of Natural Areas and Preserves

ACCESS: Open to the public during posted hours

WEB SITE RESOURCES: http://www.dnr.state.oh.us/dnap/location/lake_katharine/tabid/904/Default.aspx

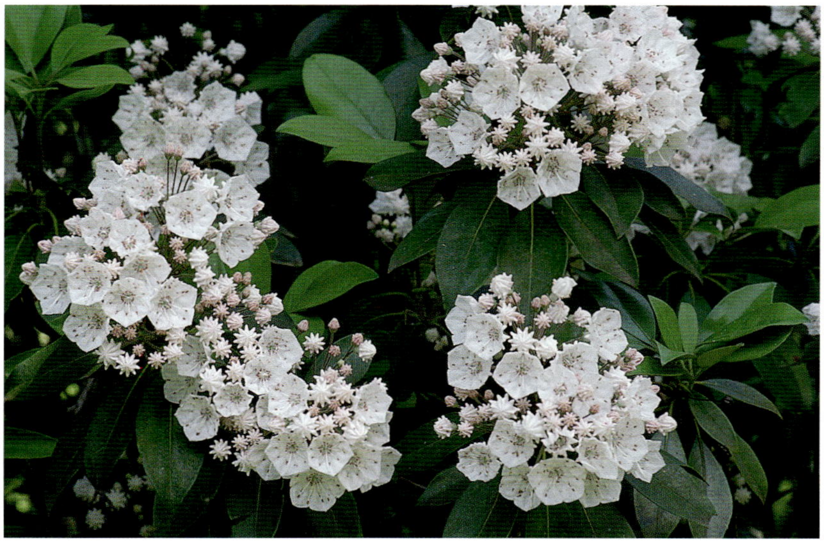

Tangles of mountain laurel often festoon the bluffs of sandstone outcrops on southern Ohio ridges. This shrub is a host plant for the rare brown elfin butterfly.

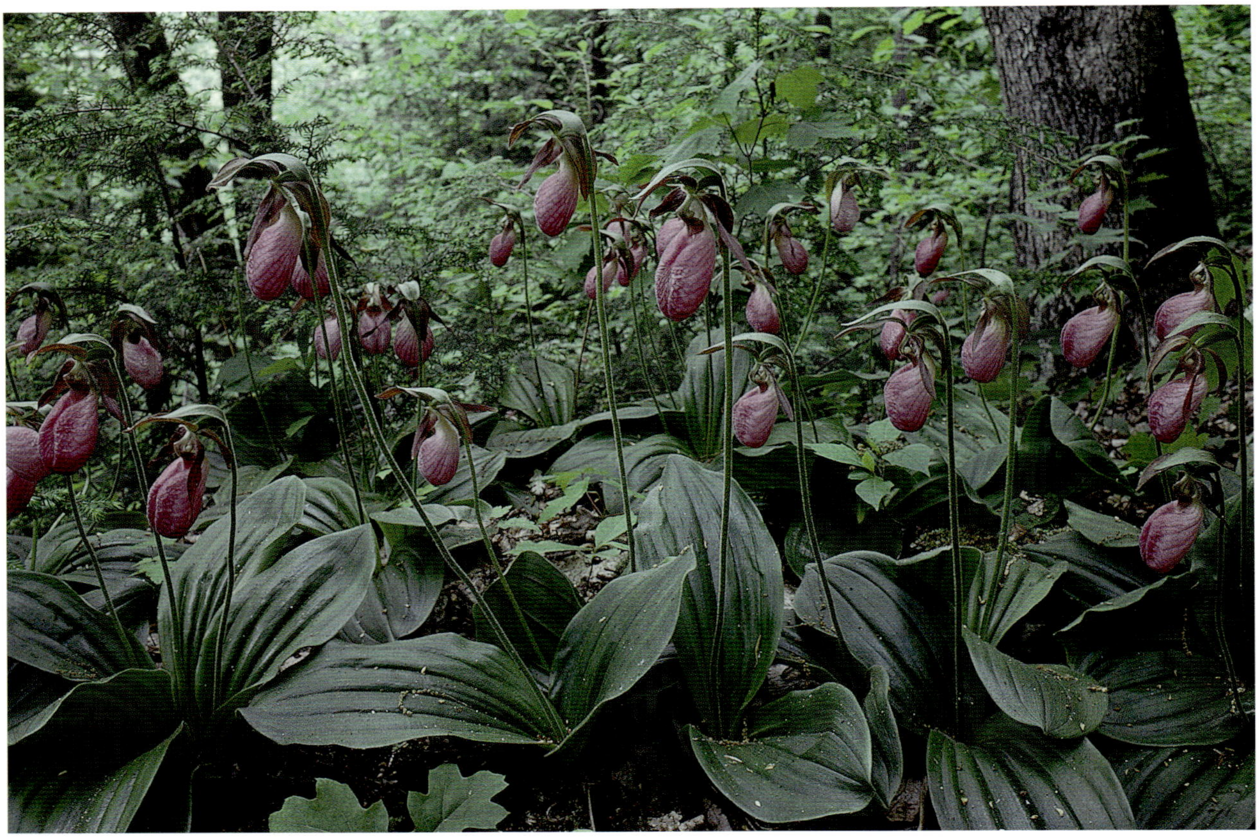

Above: Seeing a colony of exquisite pink lady's-slippers is an unforgettable spectacle. Like many orchids, lady's-slippers require specialized soil conditions involving a poorly understood relationship with mycorrhizal fungi.

Opposite: One of five native lizards in Ohio, fence lizards are lightning fast and are most often heard as they scuttle through dry leaf litter. Perseverance will reward an observer, and with patience Eastern fence lizards can be closely approached and admired.

sharp eye can spot Trudell's spleenwort, a cross between the two, and one of twenty-one hybrid ferns documented in Ohio. Far more conspicuous are bright red flowers of round-leaved catchfly, which grow in loose sand at cliff bases.

Floodplain terraces along Salt Creek and its tributaries have much richer soil than the highlands, and plant life is very different. Thick stands of pawpaw form mini-jungles along streams and produce vigorous populations of zebra swallowtails. This striking butterfly depends on pawpaw as a host plant. Towering sycamores, green ash, and tulip trees form one-hundred-foot-high canopies that shelter riparian birds such as Northern parula, yellow-throated warbler, Baltimore oriole, and cedar waxwing. About half of Ohio's 180 nesting bird species breed at Lake Katharine.

The most significant feature of the preserve may be its magnolias. Four species of these mostly southern trees occur in Ohio, and only the tulip tree is common and widespread. Rare umbrella magnolias grow on stream terraces, startling visitors with their massive leaves. Jackson County is as far north as they get. Even more amazing, and much rarer, are endangered bigleaf magnolias. The leaves of this small tree can eclipse two feet in length and create an almost jarring tropical

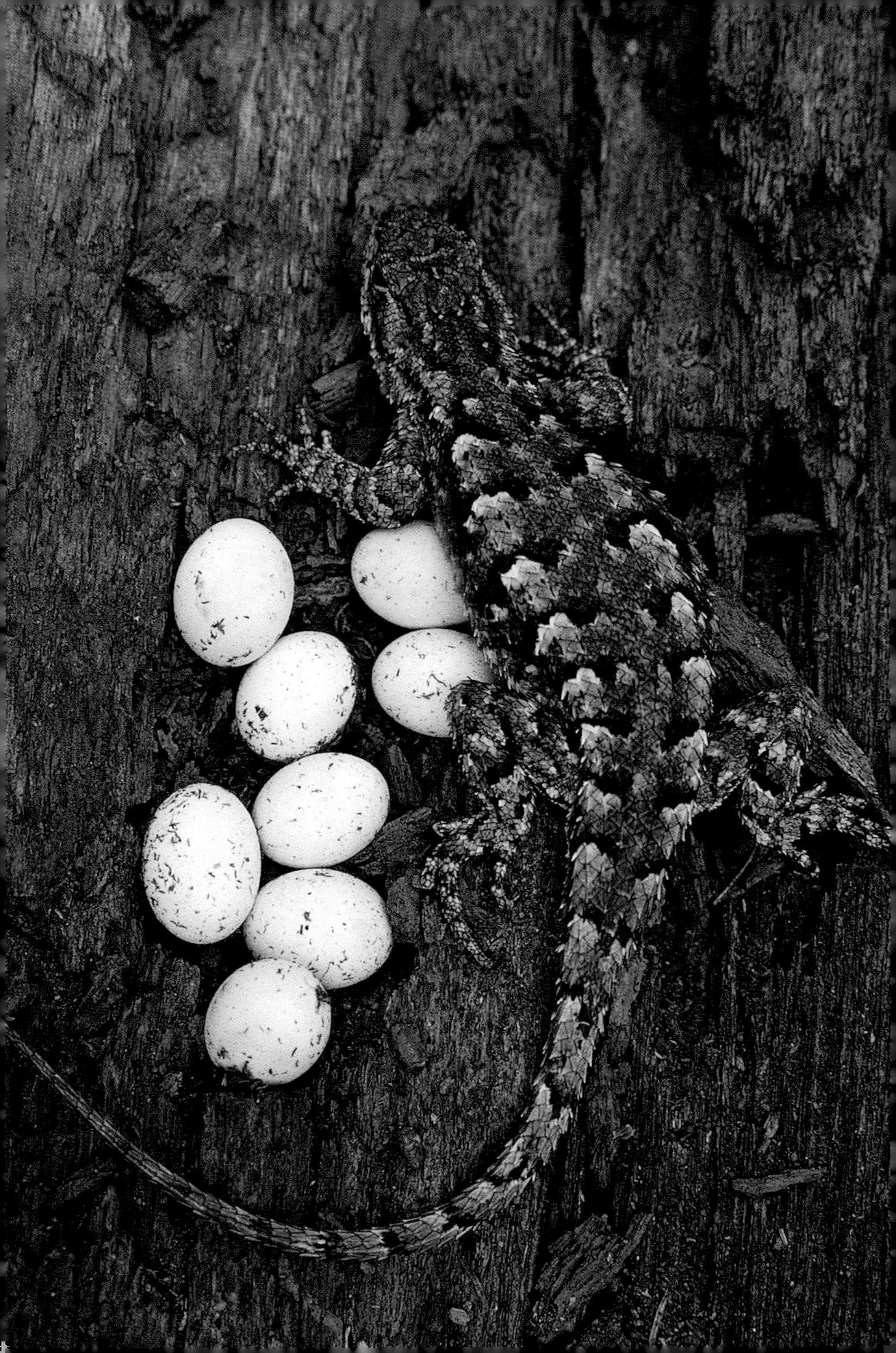

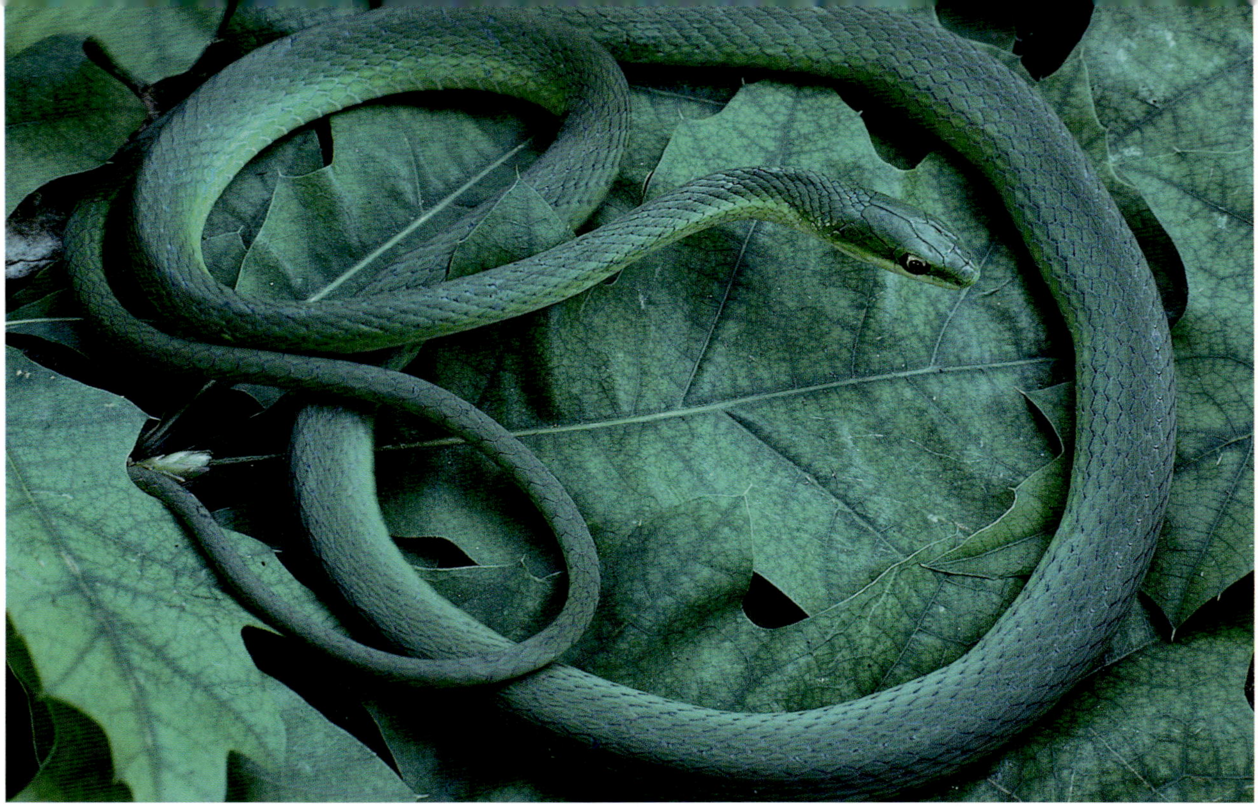

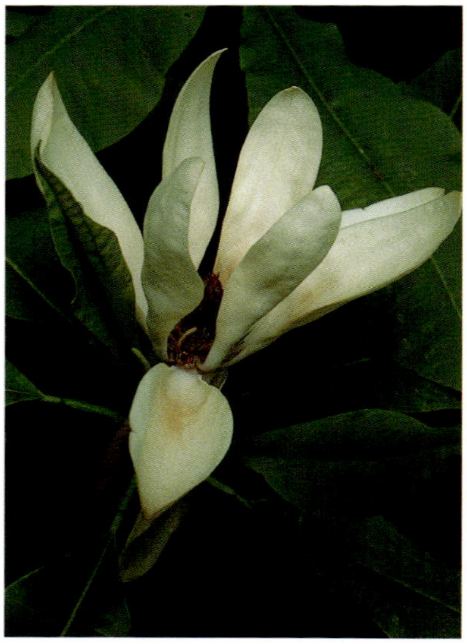

flavor in otherwise typical Appalachian woodlands. Bigleafs are only known from Lake Katharine and the immediate vicinity.

A long-dead river known as the Teays delivered magnolias and other southern flora to Lake Katharine. This stream flowed during the Tertiary period, which commenced sixty-five million years ago. Glaciation obliterated the Teays River about two million years ago. The Teays had its headwaters near what is now the Great Smoky Mountains National Park in North Carolina and ran north, entering Ohio near Portsmouth and exiting near present-day Grand Lake St. Marys. Over its eons of existence, Appalachian flora—rhododendrons, magnolias, and many others—migrated northward along the Teays. Some of these plants—the northernmost outliers of their species—still remain. In a sense they were marooned by the prehistoric river.

Top: Rough green snakes are easily overlooked, as they are arboreal and spend much time hunting insects in the branches of shrubs and trees. This species is confined to Ohio's southernmost counties.

Bottom: With two-foot leaves, the umbrella magnolia nearly sets the record for the largest simple leaves of any Ohio plant. Not a large tree, it rarely reaches a height of fifty feet.

Slicing through the western Appalachian foothills of Columbiana County, Little Beaver Creek offers a taste of presettlement Ohio. Few Ohio streams are as wild in character, a fact supported by its status as a State and National Wild River.

Little Beaver's rushing waters cascade down a rapid forty-five-foot drop at Purgatory Hollow Falls, carving through Lower Freeport sandstone formations. Elsewhere in Beaver Creek State Park, promontories of sandstone jut above the valley, providing sensational scenery.

Protecting forests is integral to protecting rivers; buffering woodlands prevent silt and other pollutants from entering the stream. And forests aplenty are along the course of Little Beaver Creek, keeping its waters pristine. There is an abundance of life, both aquatic and terrestrial. Anglers enjoy smallmouth bass fishing, and birders will be dazzled by an incredible diversity of avifauna.

The melodies of feathered songsters fill summer mornings: species such as black-and-white warbler, scarlet tanager, and wood thrush are

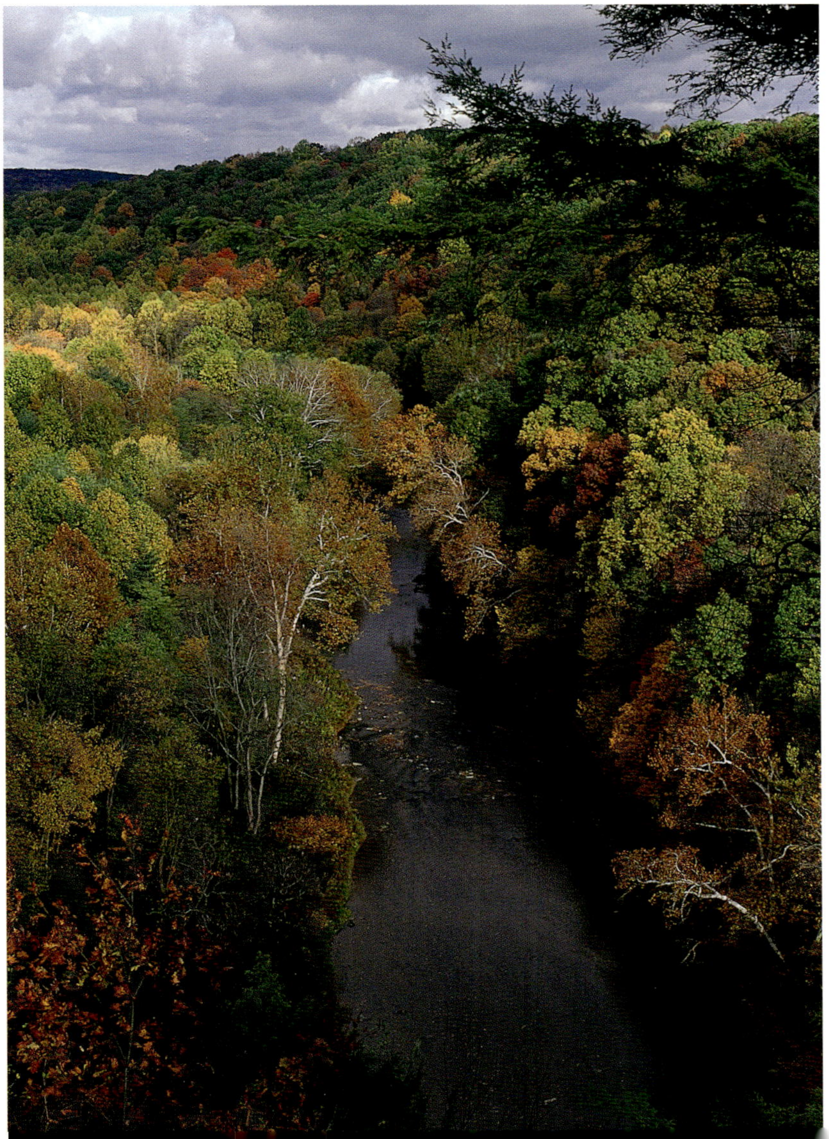

Little Beaver Creek

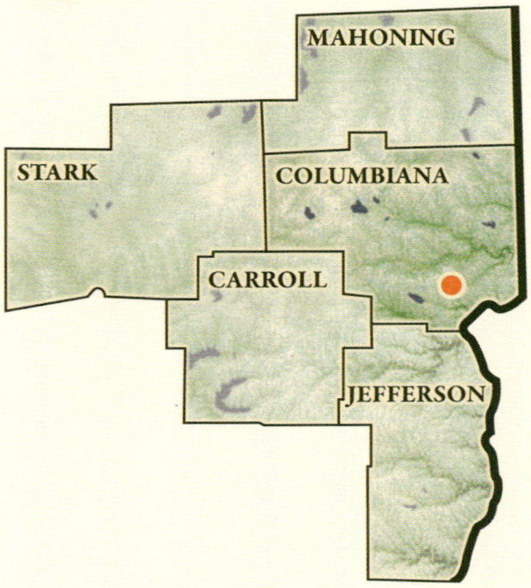

ACREAGE: ODNR lands total nearly 3,200 acres

OWNERSHIP: Miscellaneous

ACCESS: Best access via state park lands and state nature preserve. Sheepskin Hollow State Nature Preserve; Beaver Creek State Park

WEB SITE RESOURCES:
http://www.ohiodnr.com/parks/parks/beaverck.htm

http://www.ohiodnr.com/dnap/location/sheepskin.htm

http://www.ohiodnr.com/dnap/sr/lbeaver.htm

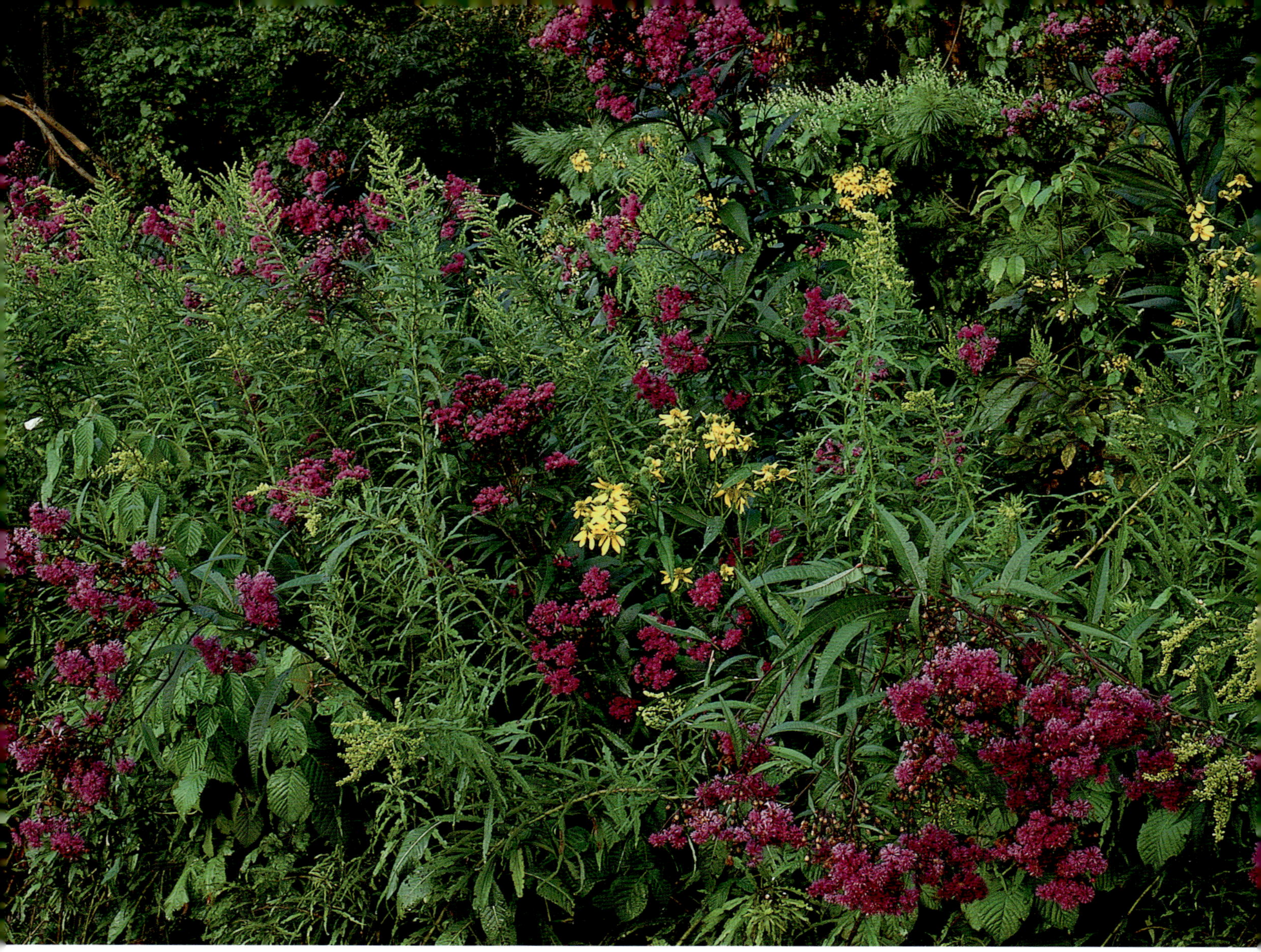

Previous page: One of the most picturesque streams in Ohio, Little Beaver is also one of the most pristine. Large forested buffers, as seen along the North Fork of Little Beaver Creek in this photo, are vital to the protection of the stream. Riparian woodlands play host to many specialized terrestrial plants and animals, and the lush cover serves as a buffer filter against contaminants. By reducing sediment and other contaminants that would otherwise enter the stream, forests ensure that water quality remains high, thus protecting the aquatic organisms.

Above: In late summer, wildflowers bloom in profusion on the banks of Beaver Creek. Unlike diminutive spring flora, autumnal herbs reach for the sky, with some species like tall iron-weed and wingstem growing head high.

common. All told, some 120 species have been documented breeding along Little Beaver. Many of them have come from far-flung haunts—the three listed above winter in Central and South America and are merely temporary visitors to our latitude.

Flora is spectacular, especially in May. Some upland oak–dominated ridges support colonies of pink lady's-slippers, an extraordinary orchid whose looks rival any tropical species. Rich slopes are carpeted with Ohio's official wildflower, large-flowered trillium, and stream terraces host blankets of gorgeous blue-eyed Marys. Accomplished botanists have fun seeking rarities like speckled wood-lily, mountain-fringe, and rock skullcap.

An important modern-day discovery was documentation of common mergansers nesting along Little Beaver—the first Ohio record. Breeding had been suspected for several years, but confirmation came in 2006 when an adult was found with a brood of young chicks. These large fish-eating ducks require clear, swift rivers with ample forests, and their presence along Little Beaver is a symbol of the wildness of this river.

Right: The spritely Northern parula is our smallest warbler, weighing but eight grams—same as a fifty-cent piece—and measuring 4½ inches in length. A neotropical migrant, parulas winter in the Caribbean and in Central America. Protecting mature riparian habitats such as are found along Little Beaver is imperative to conserving this tiny but amazing bird.

Overleaf: The hellbender is our largest salamander. Big ones can be nearly two feet in length. Little Beaver Creek harbors one of Ohio's few remaining populations of this endangered species.

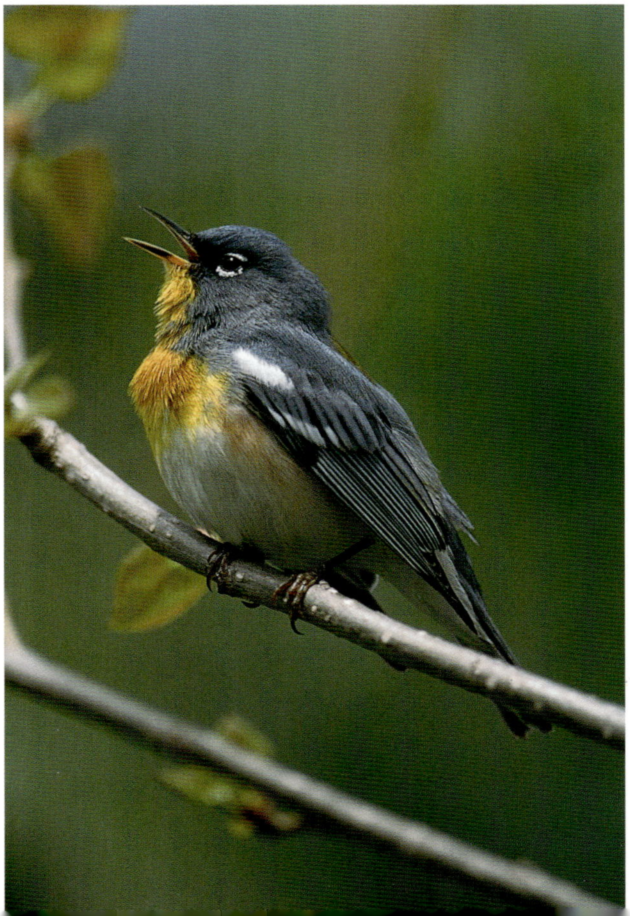

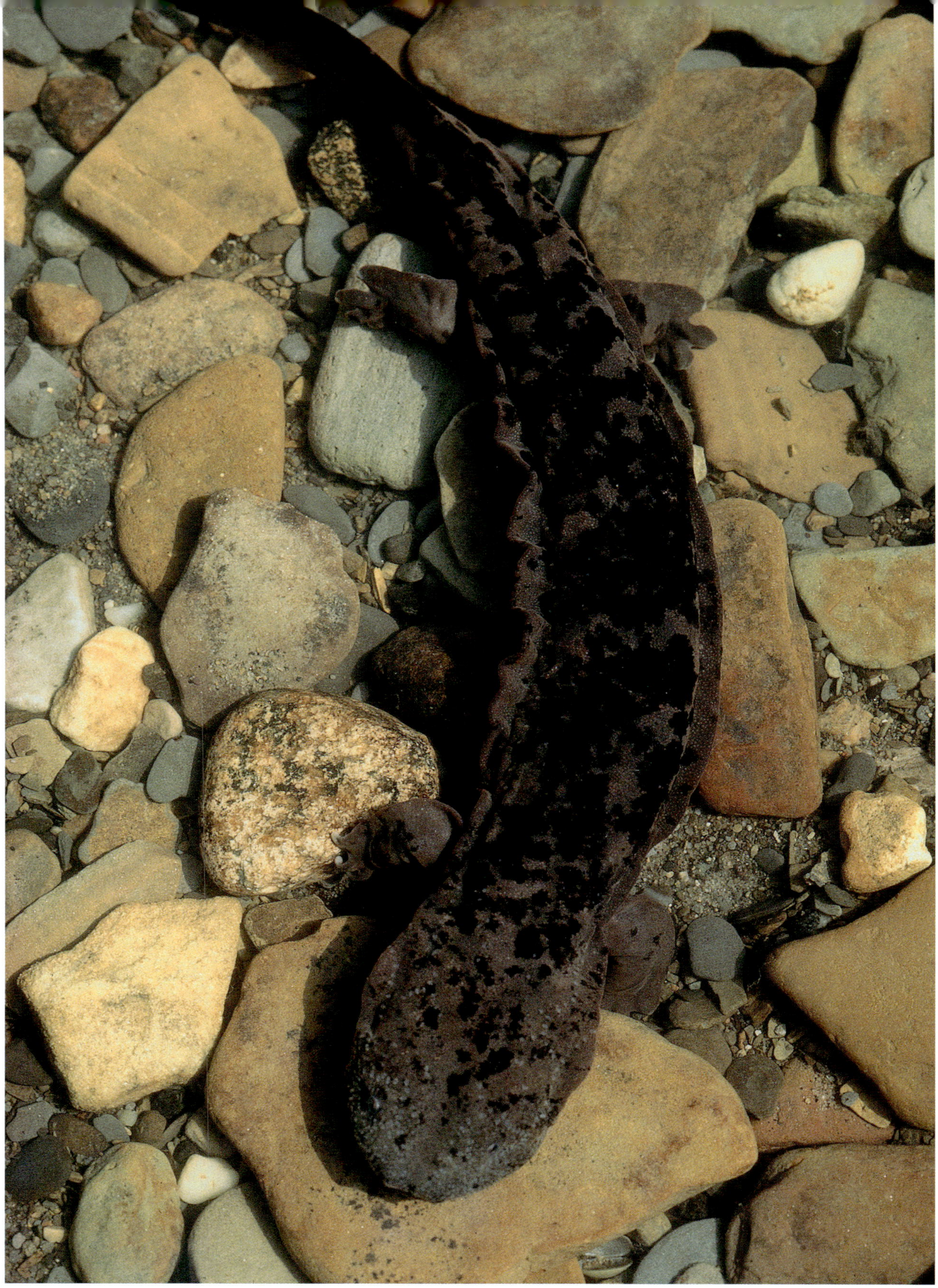

Clear Fork Gorge is the centerpiece of Mohican, a three-hundred-foot-deep gash carved through Mississippian-age sandstones. Massive ice sheets of the Wisconsin glacier skidded to a stop just a mile northward and glacial meltwaters flumed through this valley, excavating the steep-sided canyon. The two sides of the gorge differ dramatically. North-facing slopes are shaded and much cooler, with thick stands of hemlock shrouding the steep hills. Other botanical northerners include Canada yew, yellow birch, and white pine. Craggy sandstone cliffs and outcrops are dominant features, and fissures vent fifty-six-degree air from subterranean pockets, further chilling the valley. It might be ten or more degrees cooler in the valley's depths than in the uplands on a hot summer day. These cool valleys were given the name Neotomas.

Mohican ranks tenth largest among Ohio's twenty state forests, but in spite of its relatively small size it may harbor more breeding bird diversity than any other Ohio woodland. The forest is a rich tapestry of plant communities, including mature hemlock-covered slopes. Although only five thousand acres, Mohican State Forest supports over one hundred species of breeding birds.

Deep hemlock forests on the gorge's north slope support some of Ohio's rarest breeding birds. Fiery-throated Blackburnian warblers deliver high-pitched songs, inaudible to those of less-than-perfect hearing, and Canada warblers shout their short explosive warbles. Eclipsing all in sheer melodic descant are the speckled, shy, and rarely seen hermit thrushes,whose rich tunes are among nature's most extraordinary lyrics. All of these birds are abundant nesters in the expanse of boreal forest that sweeps across northern North America—a forest ecosystem greater in scope than the jungles of the Amazonian basin. At latitudes this far south, they are rare disjuncts confined to remnant boreal forests like Mohican.

South-facing hillsides of the Clear Fork gorge are warmer and dominated by mixed-mesophytic forests. Stream terraces support massive sycamores, and as one advances upslope, trees shift between tulip tree, white ash, red oak, white oak, hickory, and others. In contrast to the northerners on the opposite slope, birds on south slopes include southerners near their northern limits. Dry monotonic trills of worm-eating warblers mix with the bold ringing *churee churee churee* of Kentucky warblers. Inquisitive Carolina chickadees search boughs and cavities for prey, emitting four-noted *fee-bee fee-bay* songs. Mohican is their northern boundary—black-capped chickadees occupy the lands beyond.

Mohican State Forest

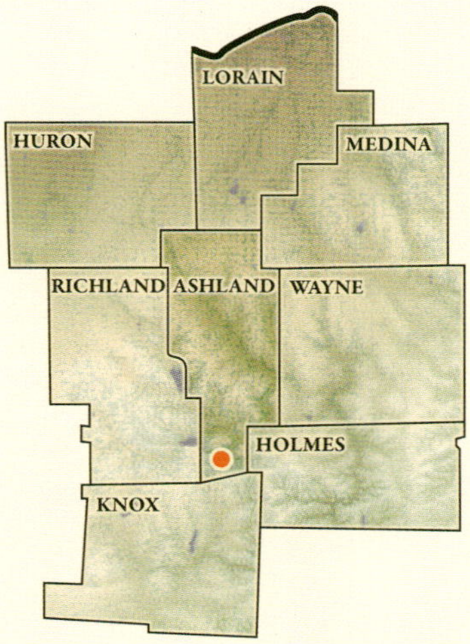

ACREAGE: 5,664 (4,525 state forest; 1,110 state park; 29 state nature preserve)

OWNERSHIP: Ohio Divisions of Forestry, Parks, and Natural Areas and Preserves

ACCESS: Open to the public during posted hours

WEB SITE RESOURCES:

http://www.ohiodnr.com/forests/mohican/tabid/5160/Default.aspx

http://www.ohiodnr.com/parks/mohican/tabid/769/Default.aspx

http://www.ohiodnr.com/dnap/location/clear_fork/tabid/928/Default.aspx

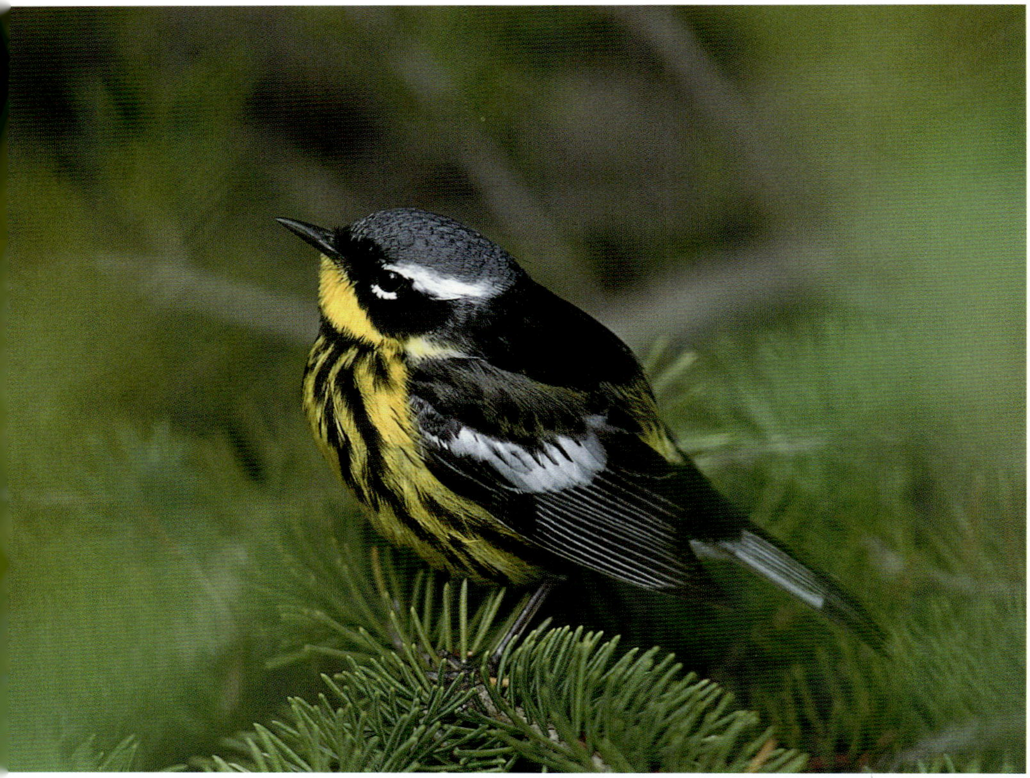

Left: Common migrants, scores of magnolia warblers can be seen on a good spring or fall migration day. Not so for breeders—they are at the southern end of their boreal breeding range in Ohio and only a few isolated populations occur.

A little-seen aspect of Mohican biodiversity is bats. These nocturnal mammals are abundant in the gorge, holing up during daylight hours under exfoliating tree bark and in rocky crevasses. At night, they emerge to hunt nighttime insects over the waters of the Mohican River. At least six of Ohio's eleven bat species have been found in the Clear Fork gorge. A common one is the Northern myotis. Possessed of Vulcanlike ears à la Mr. Spock, these bats often glean insects from foliage rather than engaging in the fabulous aerial pursuits of their counterparts. Local researchers have been tagging bats in the gorge for years, and a Northern myotis recaptured in May 2007 had been tagged in July 2002. Extraordinary creatures, bats play vital roles in insect control and are an underappreciated constituency of Ohio's natural diversity.

Because of its close proximity to major urban centers, Mohican is especially valuable to people wishing to see wild lands close at hand. This forest is an irreplaceable ecological treasure invaluable for education, research, conservation of species diversity, and enjoyment of the outdoors.

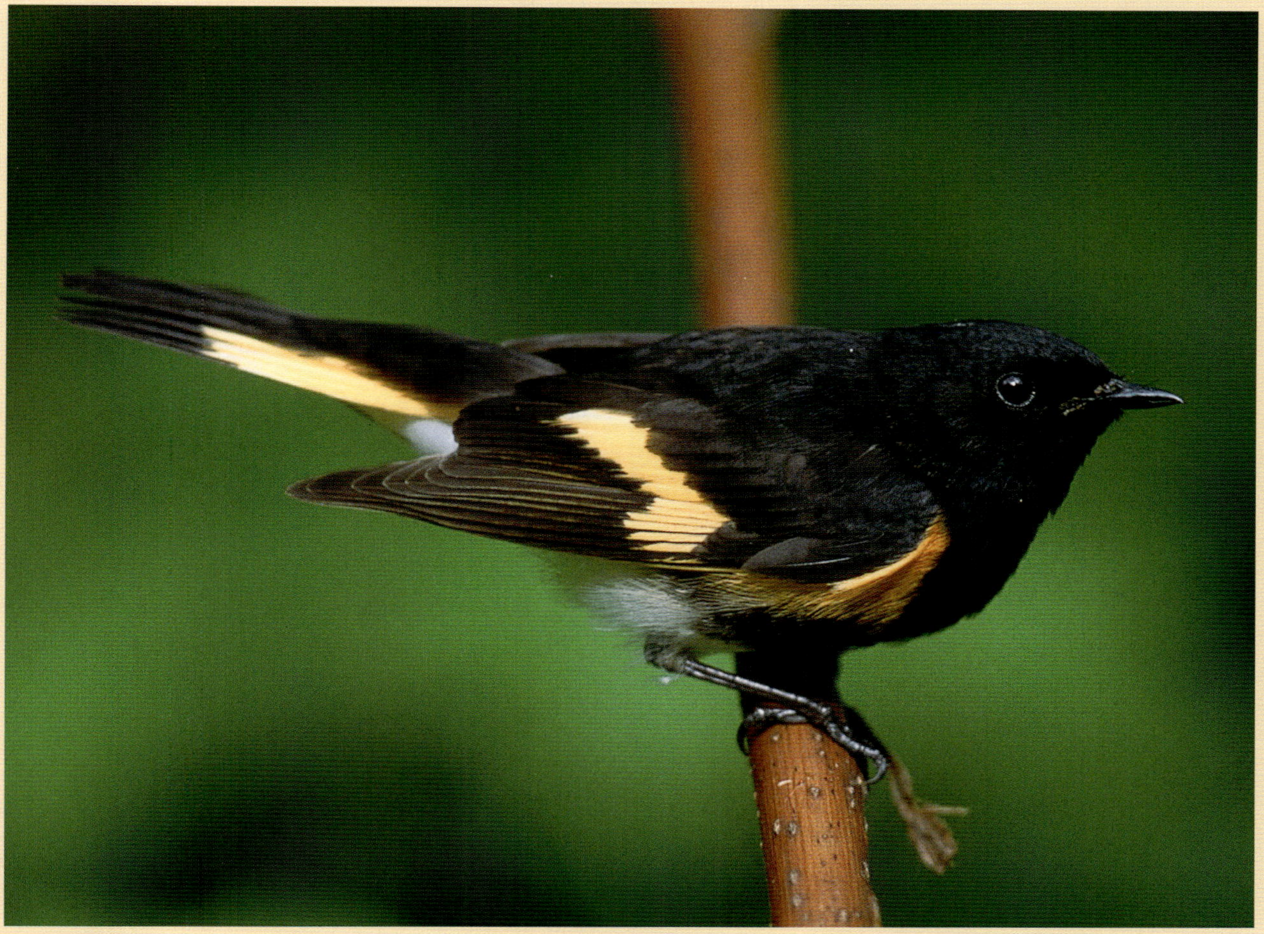

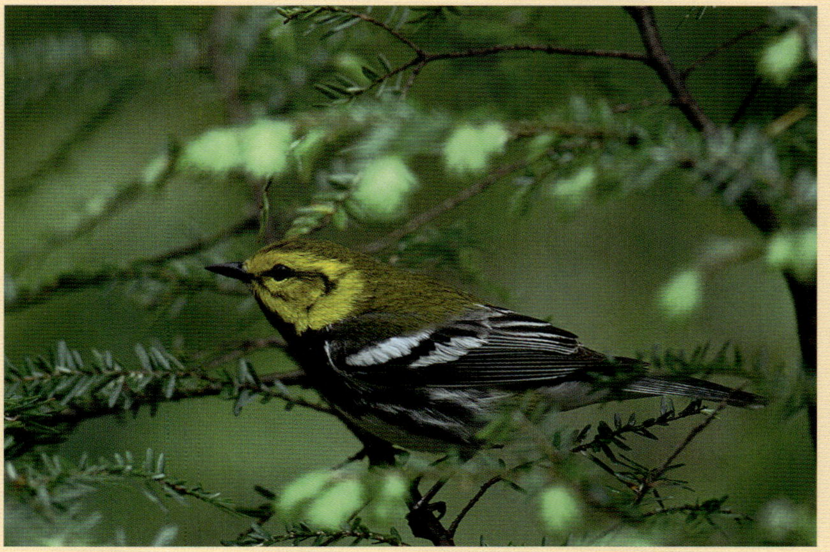

Above: American redstarts forage actively in the forest understory. The brilliant flashes of orange displayed from their constantly flicking wings and tail serve to flush insects from the foliage.

Left: An early spring migrant, the wheezy *zee zee zee zoo zee* song of the black-throated green warbler becomes a common sound by mid-April. While an abundant migrant, breeders are much scarcer and largely confined to our biggest and best hemlock gorges such as Mohican.

Overleaf: Mohican is one of the few Ohio regions in which white pine grows naturally. The trees pictured in this grove are over two hundred years old. The state champion is 153 feet tall and nearly nine feet in circumference at breast height.

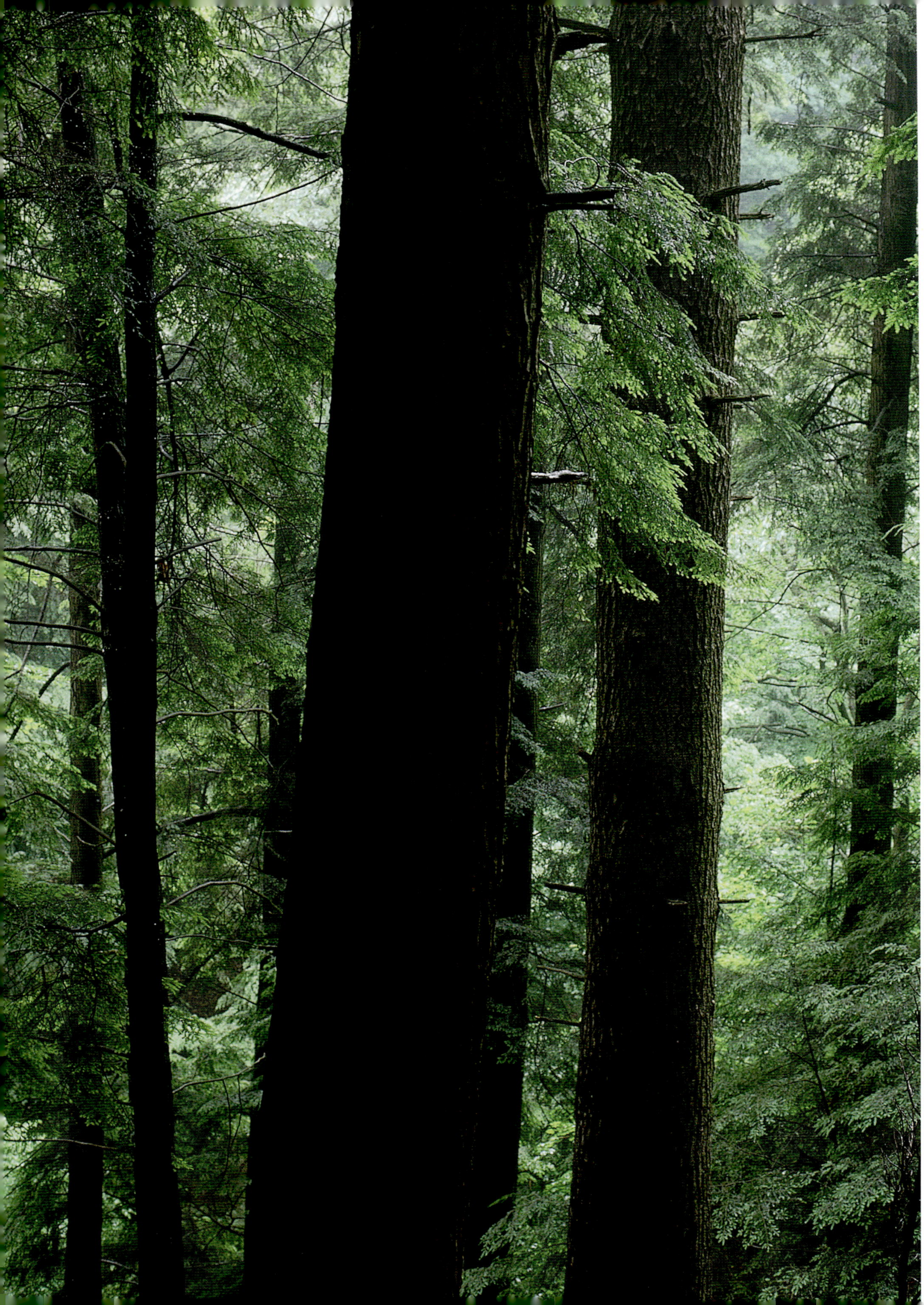

Shawnee State Forest is Ohio's greatest remaining wilderness. Seemingly endless woodlands stretch across jagged razorback ridges bisected with steep chasms bottomed with clear, rushing streams. Animals once common throughout the state and symbols of wild lands still occur here, such as bobcat, timber rattlesnake, and occasional black bears. Miles of sparsely traveled lanes traverse the woodlands, and cars are only seldom encountered.

Acquisition by the State of Ohio began in 1922, and Shawnee was a different place then. Many of the steep slopes had been cleared for small farms, and forests were grazed by farm animals. Lack of modern forestry practices caused severe erosion; many hillsides were devoid of timber, and chiseled gulches pocked barren hillsides. It's hard to imagine such lunar landscapes when gazing on today's Shawnee, carpeted with mature trees of many species filled with songs of woodland birds.

The botanical diversity of Shawnee is tremendous, possibly unrivaled elsewhere in Ohio. About one thousand species of native plants grow here, of the 1,800+ known statewide. The underlying geology strongly influences plant communities. Devonian- and Mississippian-age bedrock form the mantle on which the forest sets, and high, dry ridges are acidic and shaley. Chestnut and scarlet oaks dominate and overgrow huckleberries and blueberries. As one moves downslope, tree associations progressively become more mesophytic, and tulip tree, sweet gum, and white basswood dominate stream terraces.

Our mightiest river, the Ohio, forms Shawnee's southern boundary. The river valley has its own microclimate. Wintertime temperatures remain milder than over the hills just to the north. Formed approximately 120,000 years ago, the Ohio serves as a conduit for southern plants to migrate northward. Some plants reach their northern limits at Shawnee. Among them is American mistletoe, the same

Shawnee State Forest

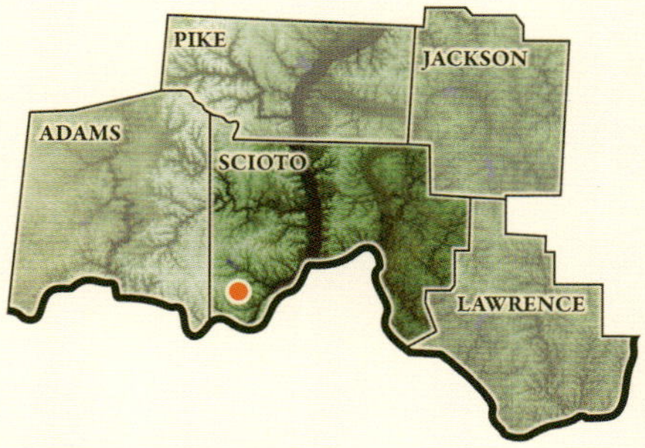

ACREAGE: 65,000

OWNERSHIP: Ohio Division of Forestry

ACCESS: Open to the public during posted hours

WEB SITE RESOURCES:
http://www.ohiodnr.com/forests/shawnee/tabid/5166/Default.aspx

http://www.ohiodnr.com/parks/shawnee/tabid/788/Default.aspx

http://www.ohiobirds.org/birdingsites/showsite.php?Site_ID=16

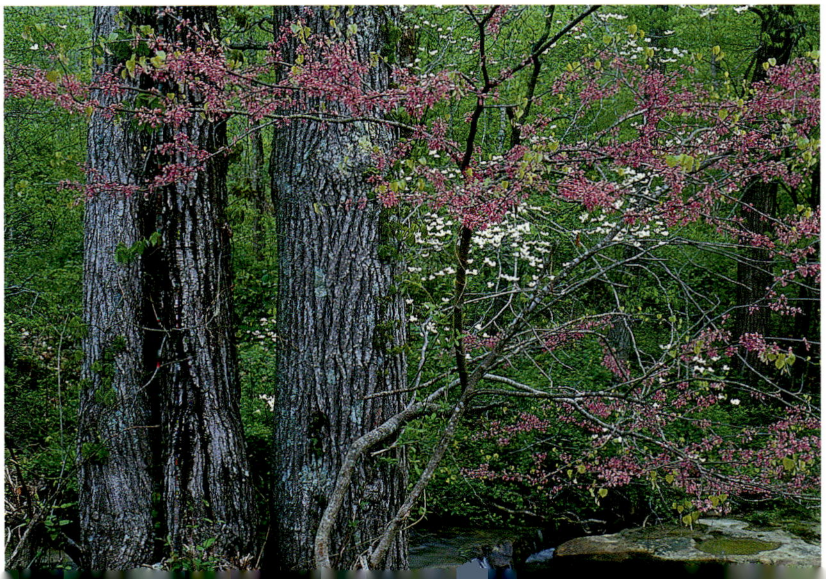

In April, redbud and flowering dogwood blushes the woodland edges of Shawnee State Forest.

species we hang in doorways at Yule. Orbicular masses of this evergreen parasite festoon riverside American elms and silver maples.

Another stream, far older than the Ohio, made Shawnee a botanical treasure trove. The Teays River, which capitulated to glacial advances of Pleistocene ice sheets two million years ago, flowed north from the Carolinas and into Ohio at Portsmouth. Over eons it transported Appalachian plants along its channel, and some still occur here, botanical relicts of glacial history. Examples include golden-star, small-flowered scorpion-weed, gall-of-the-earth, wedge-leaved violet, Southern monkshood, and early stoneroot. All are state-endangered, and none are found elsewhere north of the Ohio River.

Obvious evidence of Appalachian links are rhododendrons and azaleas. In May, aromatic pink flowers ornament pinxter-flower azaleas, and great rhododendron is sparingly scattered. Another southerner, umbrella magnolia, unfurls giant two-foot leaves on shady stream terraces. Rarest of all is Appalachian spiraea, a colonial shrub with showy panicles of white blooms. This federally threatened plant grows on gravel bars in nearby Scioto Brush Creek.

Below: Occurring in remote woodlands, timber rattlesnakes are a true wilderness symbol. Shawnee harbors the state's best populations.

Opposite: One of twenty-one species of Ohio salamanders, the mud salamander is among the rarest. New populations have recently been discovered within its known range.

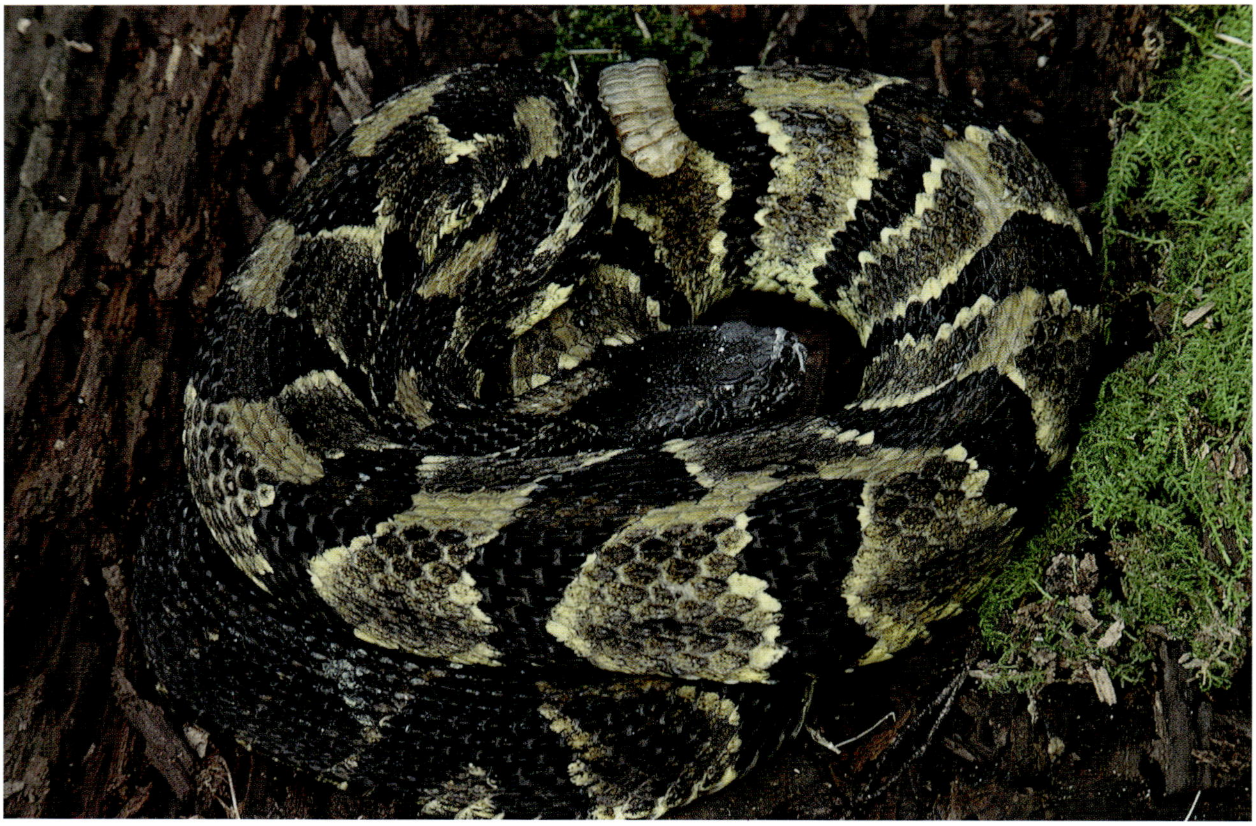

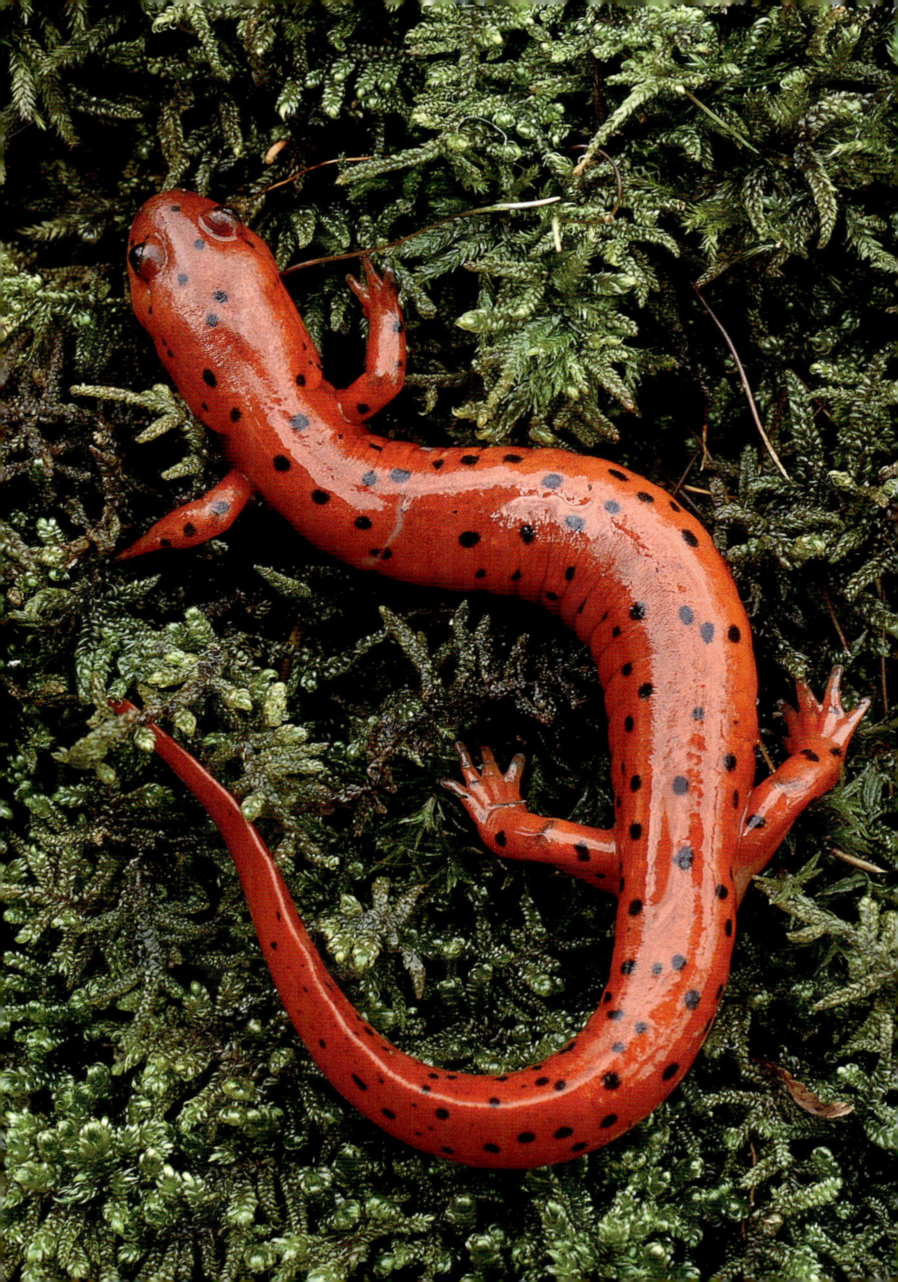

Right: Zebra swallowtails swarm in spring at Shawnee. The larvae feed on abundant pawpaw.

Below: Large numbers of worm-eating warblers find Shawnee's ravines ideal for nesting. Shawnee is one of Ohio's top birding areas.

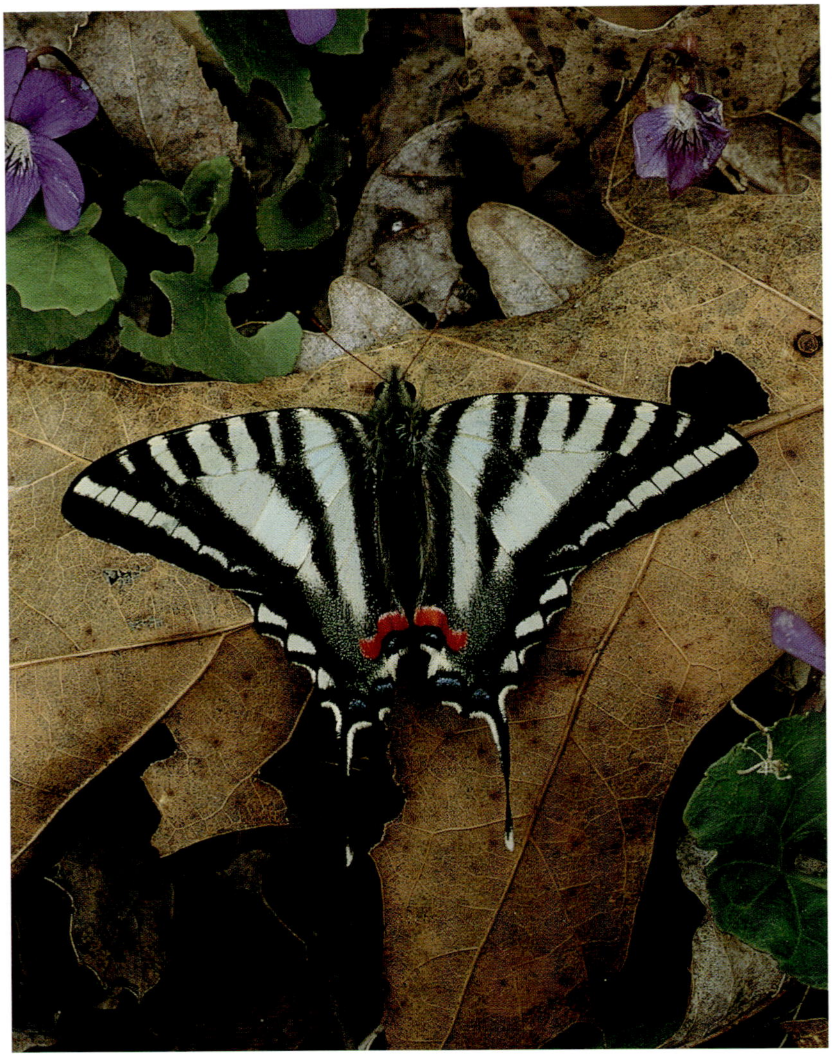

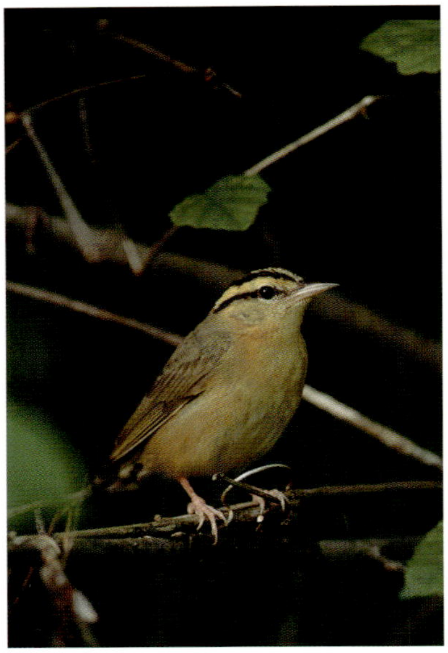

The forest is an avian paradise. Staggering numbers of woodland breeders fill the air with a riot of song in spring and summer. Cerulean warblers, a declining neotropical species, are abundant. Their buzzy spiraling songs drift from oak canopies; they're much more muted in Colombian and Venezuelan jungles where they winter. Woodpeckers abound; six species can be found nesting. Hairy and pileated woodpeckers are common; both require more extensive mature woods than do their brethren. Flycatchers of four species nest, including large, showy great crested flycatchers. Exhaling war-whoop *wheeep* calls, this cinnamon, yellow, and gray mid-canopy inhabitant is our only cavity-nesting flycatcher. Entrance holes are often adorned with a snakeskin.

Shawnee State Forest's biological significance cannot be understated. Species richness is probably unparalleled north of the Ohio River. At least forty-eight state-listed rare plants grow, including many found nowhere else in Ohio. One of only two Ohio records of the federally threatened small whorled pogonia orchid was found in Shawnee. New botanical finds are routinely made. Clear brooks have threatened rosyside dace; rare tiger spiketail dragonflies patrol overhead. Reptiles abound: four species of lizards, seventeen of snakes, and five turtle species. Twenty-nine species of amphibians have so far been tallied as well as forty-four mammals. Several dozen butterfly species help pollinate the plants. Moth diversity? No one really knows how many might be here, and that's true of most insect groups.

Birds really tie Shawnee to the global bio-economy. Over half of the state's sixty-one species of breeding neotropical birds nest here. Yellow-throated warblers travel to Caribbean islands for the winter. Ruby-throated hummingbirds journey from Shawnee to Central America, a trip that includes five hundred nonstop miles across the Gulf of Mexico. In all, birds that nest in Shawnee will disperse to ancestral habitats in Mexico and in every country of the Caribbean, Central America, and South America. How we treat Ohio habitats has an impact far beyond our borders.

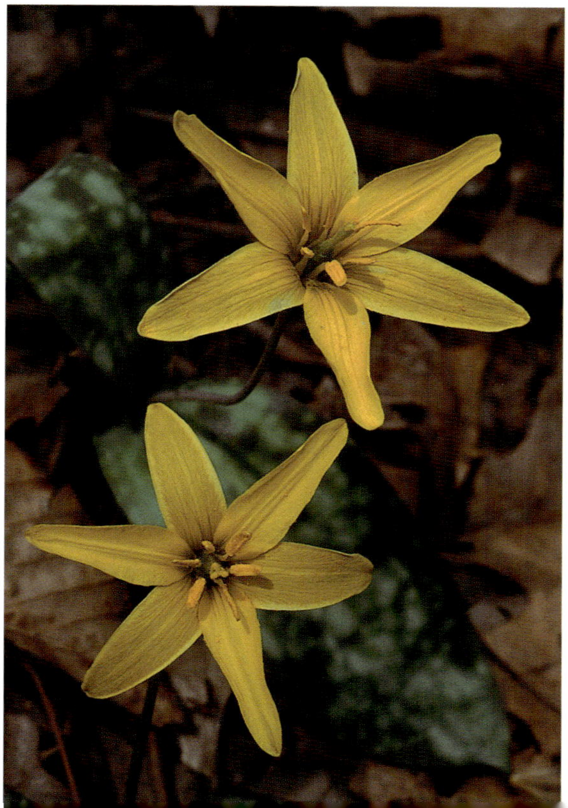

The entire golden-star population blooms within the space of a few early April days. It resembles the far commoner yellow trout-lily. The rarest lily in Ohio, it was discovered in the Rocky Fork drainage by Ohio botanist Emma Lucy Braun.

Tri-Valley Wildlife Area

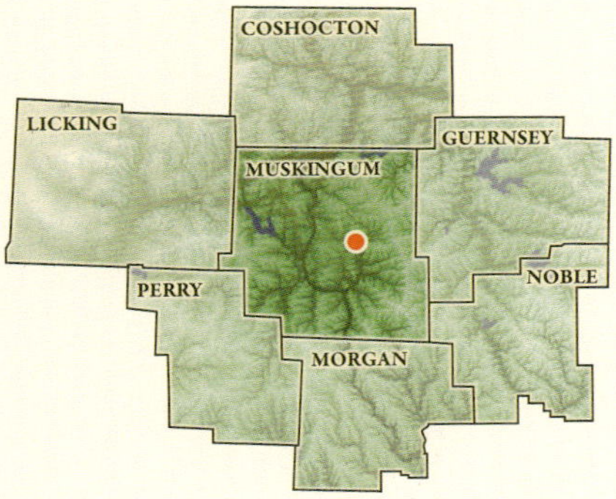

ACREAGE: 16,200

OWNERSHIP: Ohio Division of Wildlife

ACCESS: Open to the public during posted hours

WEB SITE RESOURCES:
http://www.ohiodnr.com/Home/wild_resourcessubhomepage/ild-lifeAreaMapsRepository/tabid/10579/Default.aspx

http://www.ohiobirds.org/birding-sites/showsite.php?Site_ID=93

A visitor blindfolded and plunked down unawares in the midst of Tri-Valley might think he or she landed in Africa. This sprawling grassland resembles plains where zebra and rhino roam, quite unlike anything found naturally in Ohio.

And Tri-Valley is anything but natural, and thus a big departure from every other site included in this book. But it is worthy of inclusion, as through sheer serendipity this grassland and others like it have imperfectly duplicated Ohio's former prairies, more than 99 percent of which have been destroyed. Along with the loss of native prairies came a great reduction in grassland birds that depended upon prairie landscapes.

The story of Tri-Valley is inextricably linked to coal. Massive subterranean seams of black gold underlay the hills of southeastern Ohio, and to extract it mining companies developed a technique called strip mining. Massive draglines were built to scoop away overlying topsoil, allowing the coal to be chiseled out and carted off. The downside was the destruction of hundreds of thousands of acres of fertile deciduous forest and their attendant biodiversity.

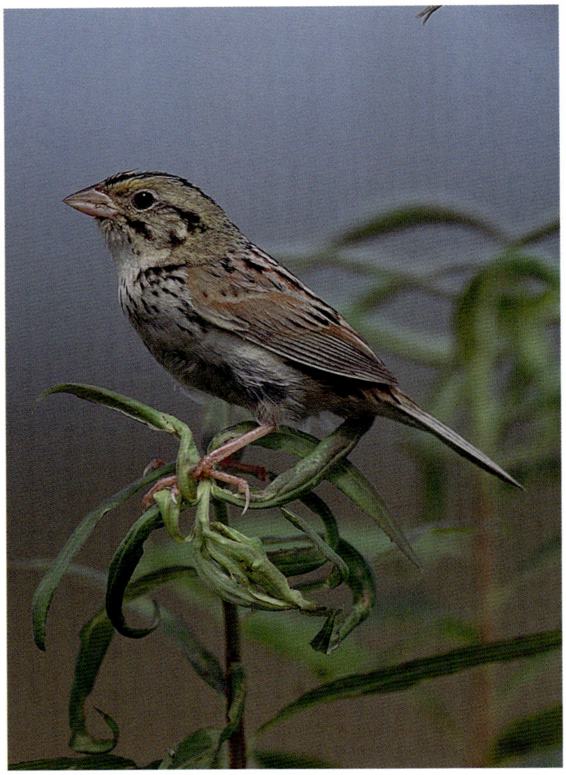

Right: Many birders make pilgrimages to reclaimed mine sites like Tri-Valley to see this secretive, declining species. Ohio sites harbor some of the largest Henslow's sparrow populations anywhere.

Opposite top: Formally dressed blackbirds, bobolinks are common breeders in the meadows at Tri-Valley.

Opposite bottom: Northern harriers are rare breeders in Ohio. Reclamation grasslands support some of the few nesting pairs of this state-endangered species.

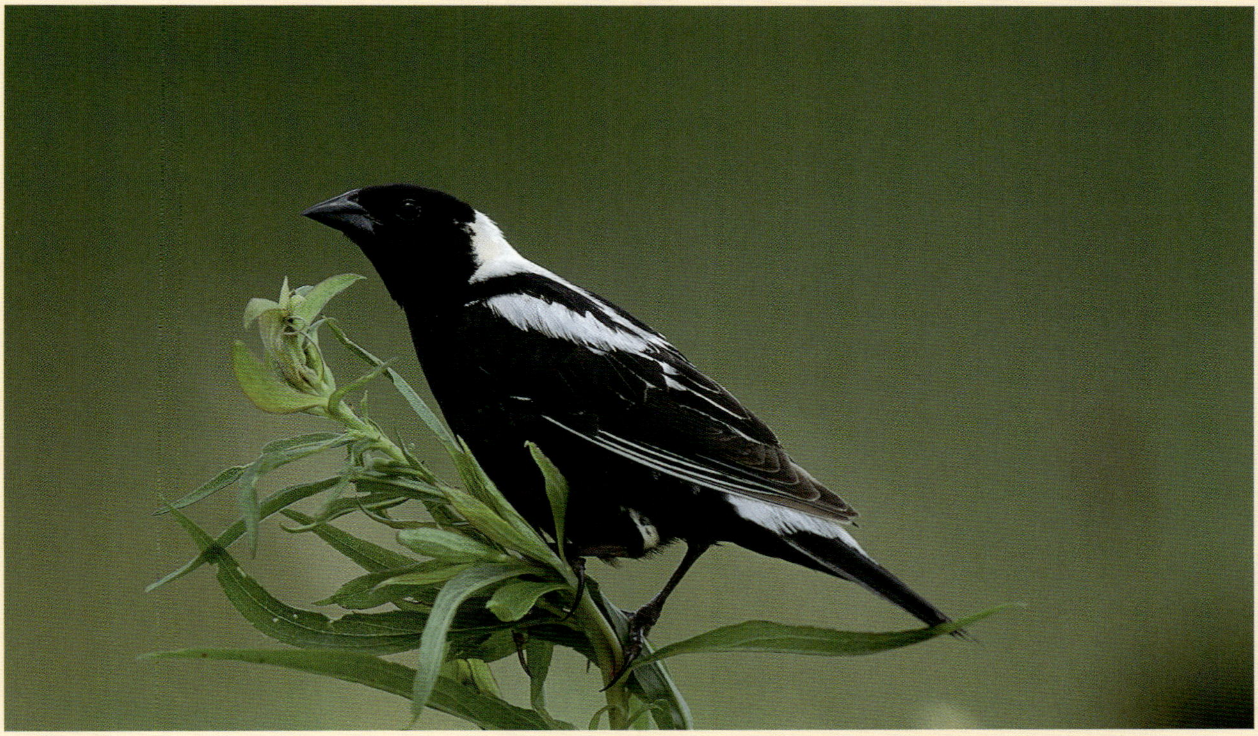
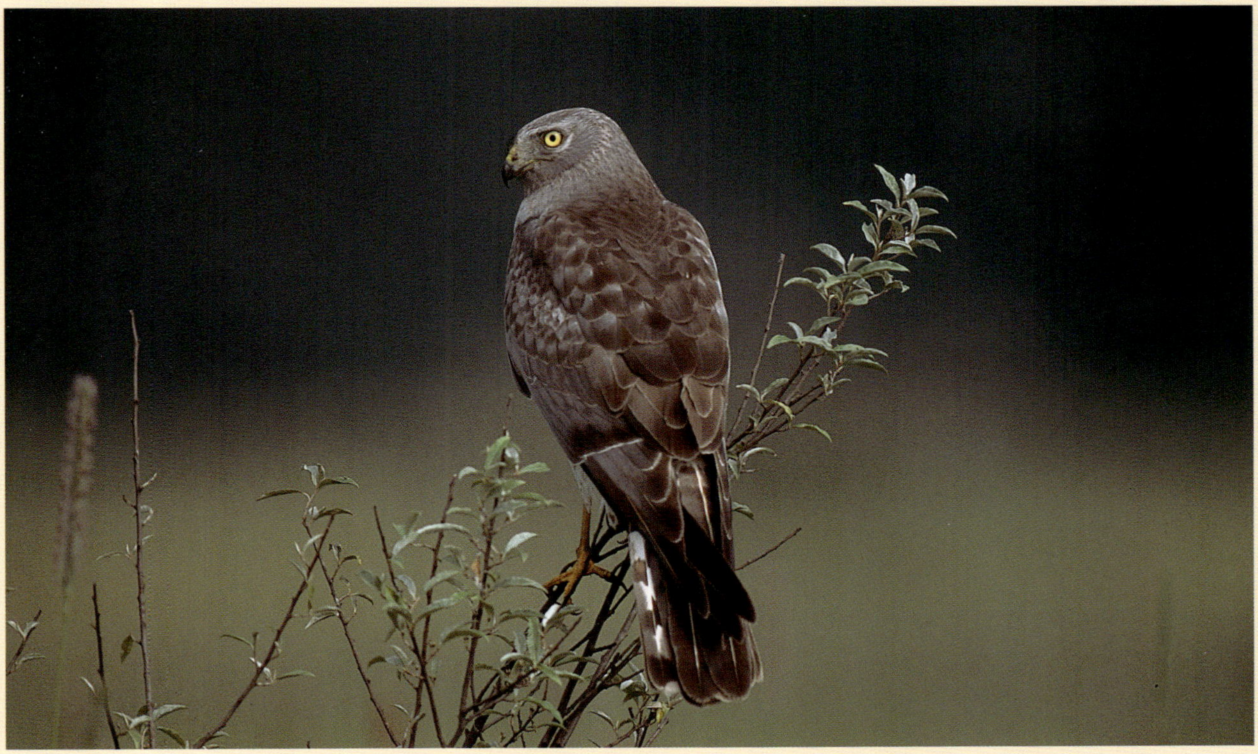

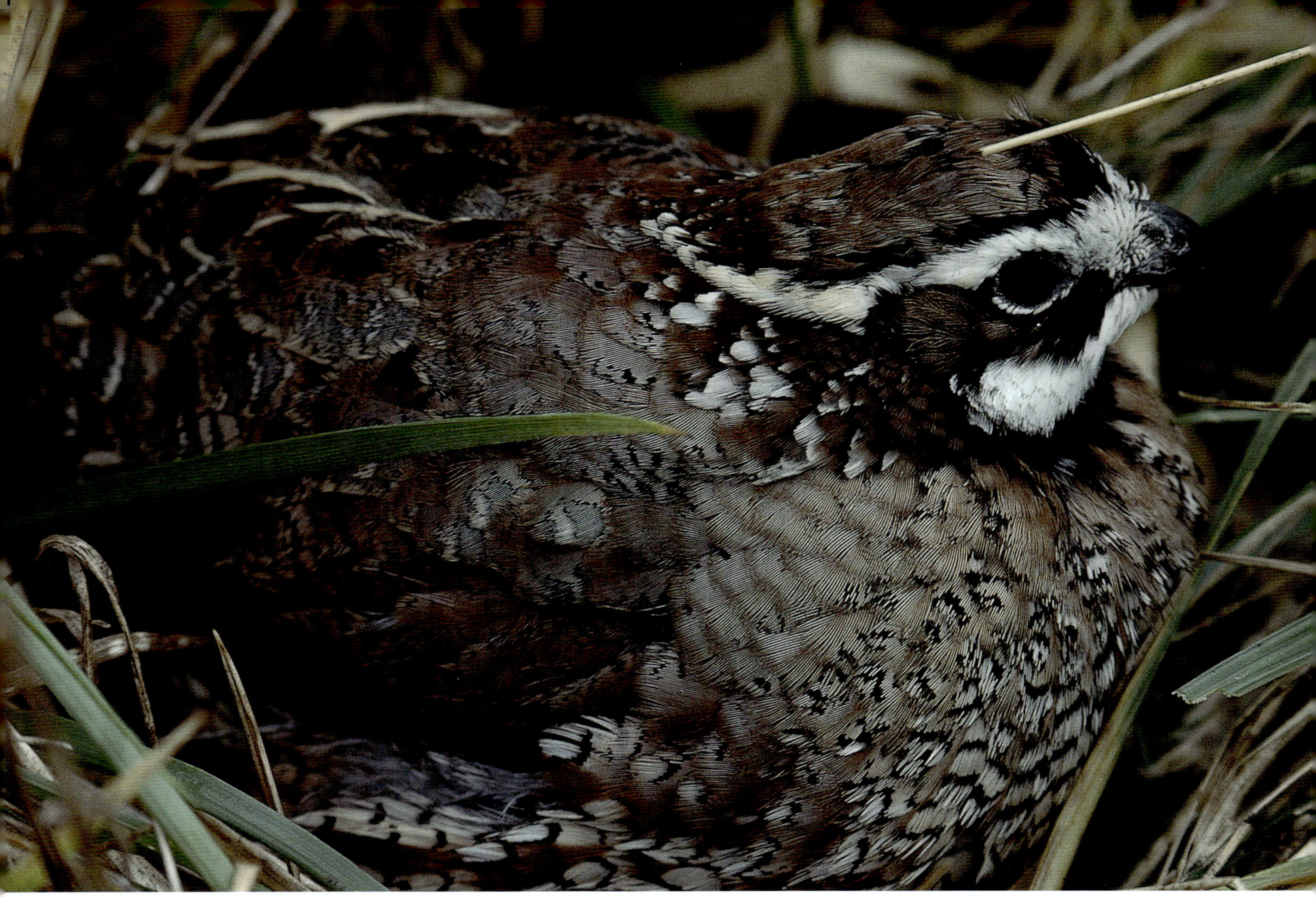

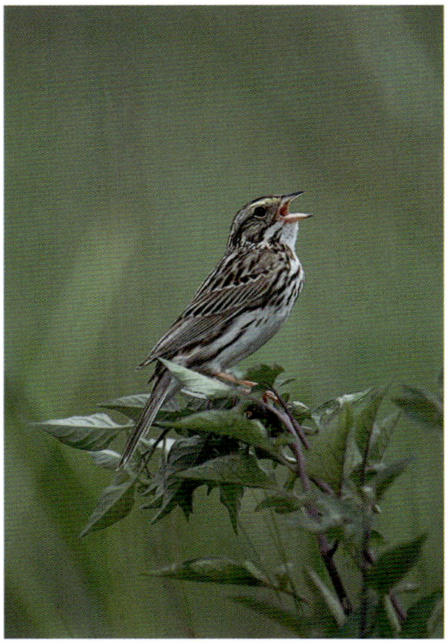

In 1972, Ohio legislators enacted House Bill 521, requiring miners to rehabilitate the lunar landscapes caused by strip mining. Reclamation involved smoothing away jagged contours of the land, and planting the newly smoothed ground to grasses to stabilize soil. Perhaps a quarter million acres of reclamation lands now sprinkle southeastern Ohio. Virtually all grasses and other plants used in reclamation are non-native. Nonetheless, many species of birds that had declined due to prairie destruction and the loss of smaller, more ecofriendly farms flourished at Tri-Valley and similar sites.

Today, birders visiting Tri-Valley in summer will be dazzled by a medley of bobolinks, dickcissels, Eastern meadowlarks, grasshopper sparrows, and many more. This site and others like it now harbor Ohio's largest and most significant concentrations of many grassland birds, including rare breeders such as the Northern harrier and short-eared owl.

Of particular interest to ornithologists is the mouselike, cryptic Henslow's sparrow. This shy, retiring bird has suffered the greatest population decline of any eastern sparrow. They have taken to reclamation grasslands like Tri-Valley with a vengeance, and these areas now contain the largest Henslow's sparrow colonies in Ohio.

Grasslands such as Tri-Valley become globally significant when one considers the bobolink, which sometimes nests here in large numbers. A colorful blackbird, bobolinks have a delightful song. Males, while on the wing, deliver a bubbly cacophony of gurgling whistles, melodies so sweet they inspired William Cullen Bryant to commemorate them in poetry. Bobolinks make long journeys to return to Muskingum County. Most winter in Argentina.

Wintering raptors exploit Tri-Valley and other reclamation grasslands. Large numbers of Northern harriers, rough-legged and red-tailed hawks, and short-eared owls converge to feast on abundant meadow voles—mobile furry sausages to the birds. Another reclamation grassland, the Wilds, has hosted golden eagle and prairie falcon—very rare wintertime visitors. Unnatural as they may be, Tri-Valley and other former strip mines still play host to important elements of Ohio's bird life.

Top left: The cheerily whistled *bob-BOB-WHITE!* was once a ubiquitous sound in Ohio's countryside. As farming practices became less bird-friendly, suitable habitat for our only native quail steadily diminished. A reintroduction program at Tri-Valley by the Ohio Division of Wildlife has been successful, and quail are easily heard here.

Bottom left: Savannah sparrows are frequent nesters at sites such as Tri-Valley, where their insectlike *tsee-tsayyy* trill is a common sound.

Interior Low Plateau—Bluegrass Section

Chaparral Prairie State Nature Preserve (40) • Edge of Appalachia Preserve (38) • Lynx Prairie (39)

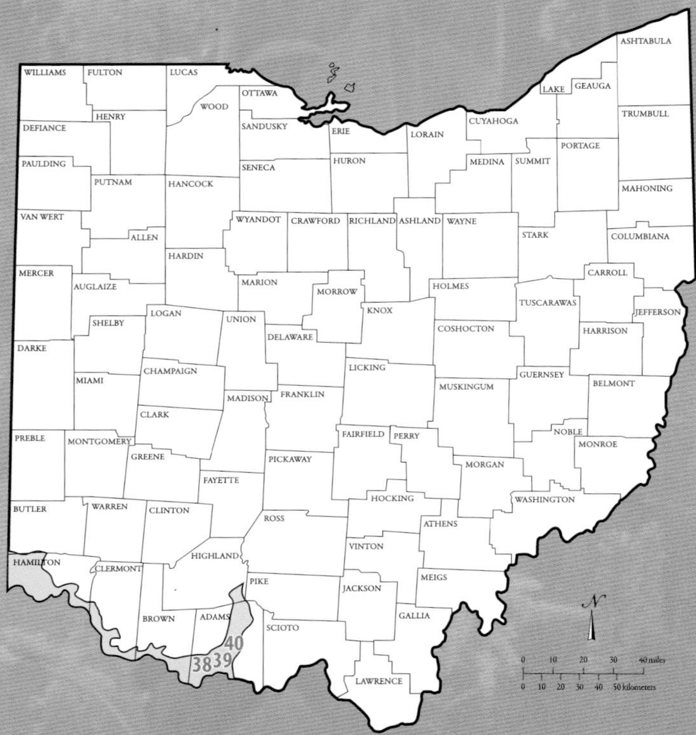

Chaparral Prairie is a plant enthusiast's dream. To the uninitiated, the prairie might look like a cedar-speckled old field, but that assessment would completely miss the mark.

Adams County prairies are far older than tallgrass prairies of the glaciated till plains. Neither of the last two glaciers scoured any but the northwest corner of the county; thus, upland soils elsewhere are thin and nutrient-poor. Stepping into Chaparral Prairie is to gaze upon a landscape that is probably much as it was tens of thousands of years ago. Because of its highly specialized prairie habitats, Adams County has more globally rare plants than any other Ohio county.

Chaparral and neighboring prairies are in our smallest physiographic province, the Bluegrass region, which barely juts into Ohio from the south. These odd prairie barrens occur on Silurian limestone, some of the oldest exposed bedrock in the state. One of the most conspicuous features of these prairies is a conifer, the red cedar. Red cedars are calciphiles (prefer neutral or above pH) and thrive on limey barrens of Chaparral, to the point that some call these habitats cedar glades.

Rare Adams County habitats have long been studied, beginning with pioneering exploration by famous Ohio botanist and ecologist Emma Lucy Braun of Cincinnati. She was the first to describe this region and document many of the rare plants found here. Braun went on to publish a classic ecological treatise in 1950, *Deciduous Forests of Eastern North America,* inspired in part by her work in Adams County.

Chaparral Prairie State Nature Preserve

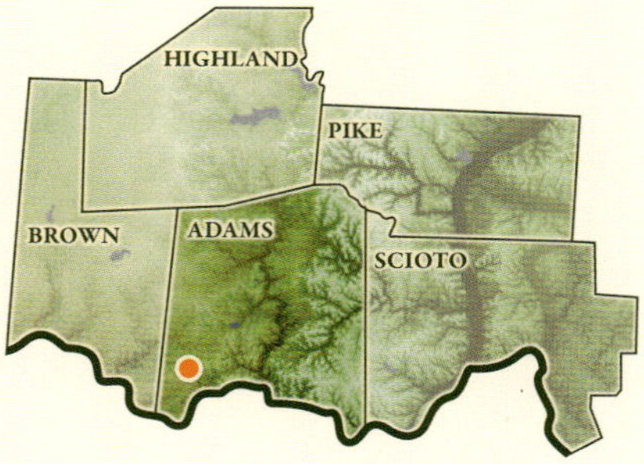

ACREAGE: 67

OWNERSHIP: Ohio Division of Natural Areas and Preserves

ACCESS: Open to the public (entrance gate may be closed; park near gate and walk in)

WEB SITE RESOURCES:

http://www.ohiodnr.com/dnap/location/chaparral_prairie.html

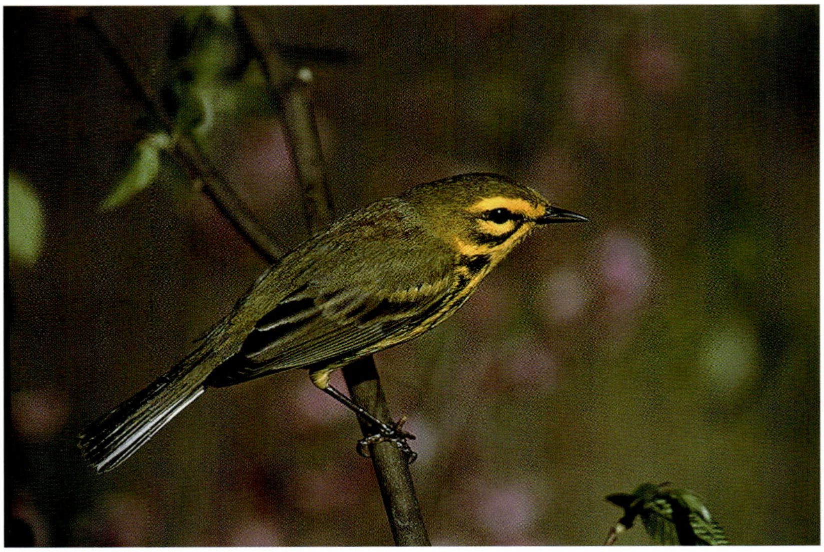

Few breeding birds spend less time in Ohio than the prairie warbler. Arriving in mid- to late April, prairies set up shop, nest, fledge their young, and depart for their tropical ancestral homes by mid-August. Adams County cedar glades support some of the best breeding populations anywhere.

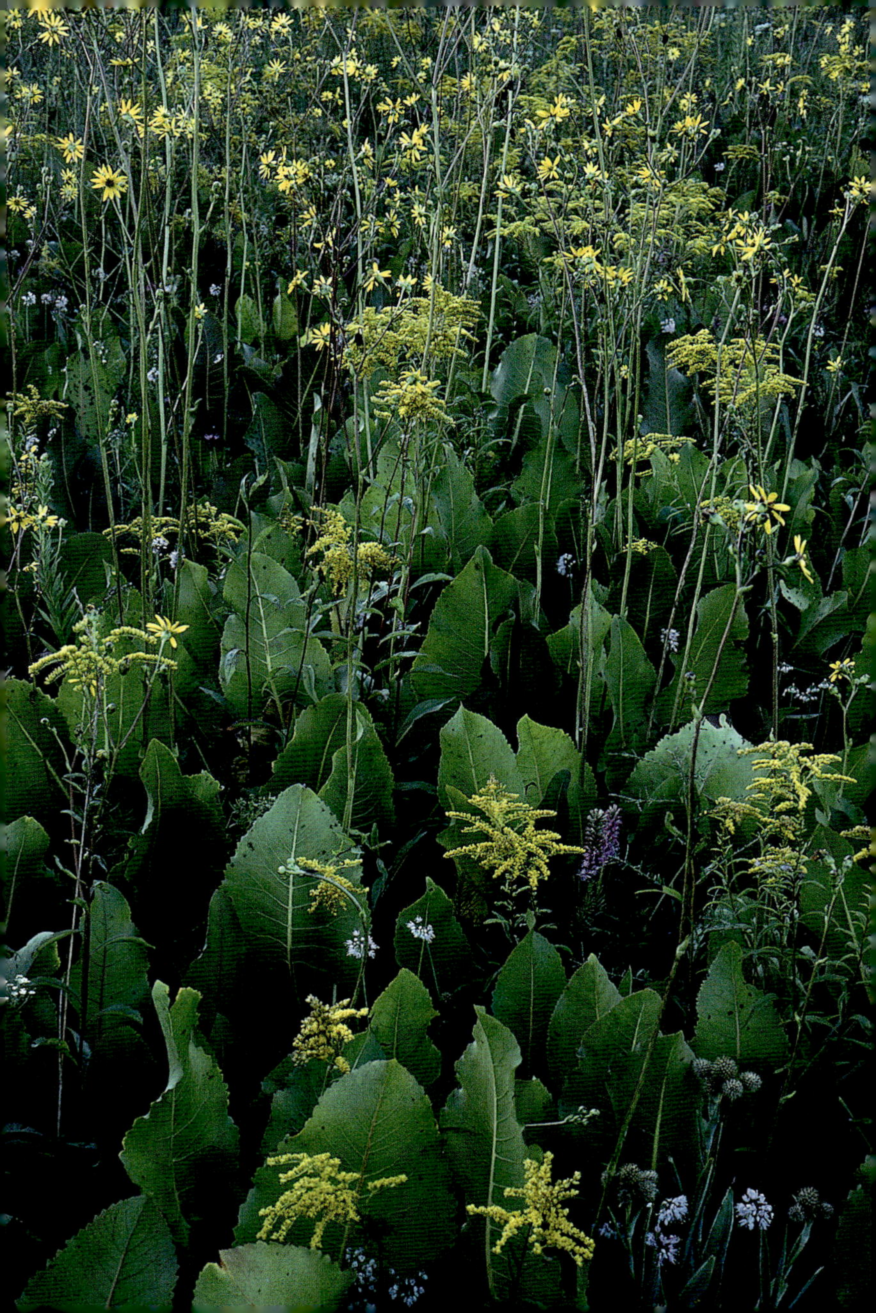

A visit to Chaparral Prairie State Nature Preserve is a wonderful introduction to the glories of an Adams County barrens prairie. Midsummer paints the arid landscape purple and yellow with spikes of blazing-star and various sunflowers, including the rare, nearly leafless Western sunflower.

The big showy plants can't be missed. Not so with some Lilliputian gems that lurk down low. In a few eroding gullies are tiny, threatened pink milkworts, only a few inches high. In the same habitat but rarer yet is Carolina leaf-flower, a miniscule spurge. State-endangered, the leaf-flower is known from only two other locales. The uncommon juniper hairstreak can also be found here, as well as in other Adams County cedar glades. It is our only predominately green butterfly.

Beginning in the late 1980s the Ohio Division of Natural Areas and Preserves began to conduct controlled burns at Chaparral. The results have been spectacular, reducing red cedars and opening up the prairie to sunlight, allowing rare plants like rattlesnake-master, prairie false indigo, and blueheart to proliferate.

Opposite: As summer progresses, tall perennials like the prairie-dock may reach eight feet or more skyward. This plant is a good prairie indicator species.

Below: Quite the rarity, bluehearts are listed as threatened in Ohio and are only known from a handful of sites. Controlled burning is important to the successful conservation of this plant.

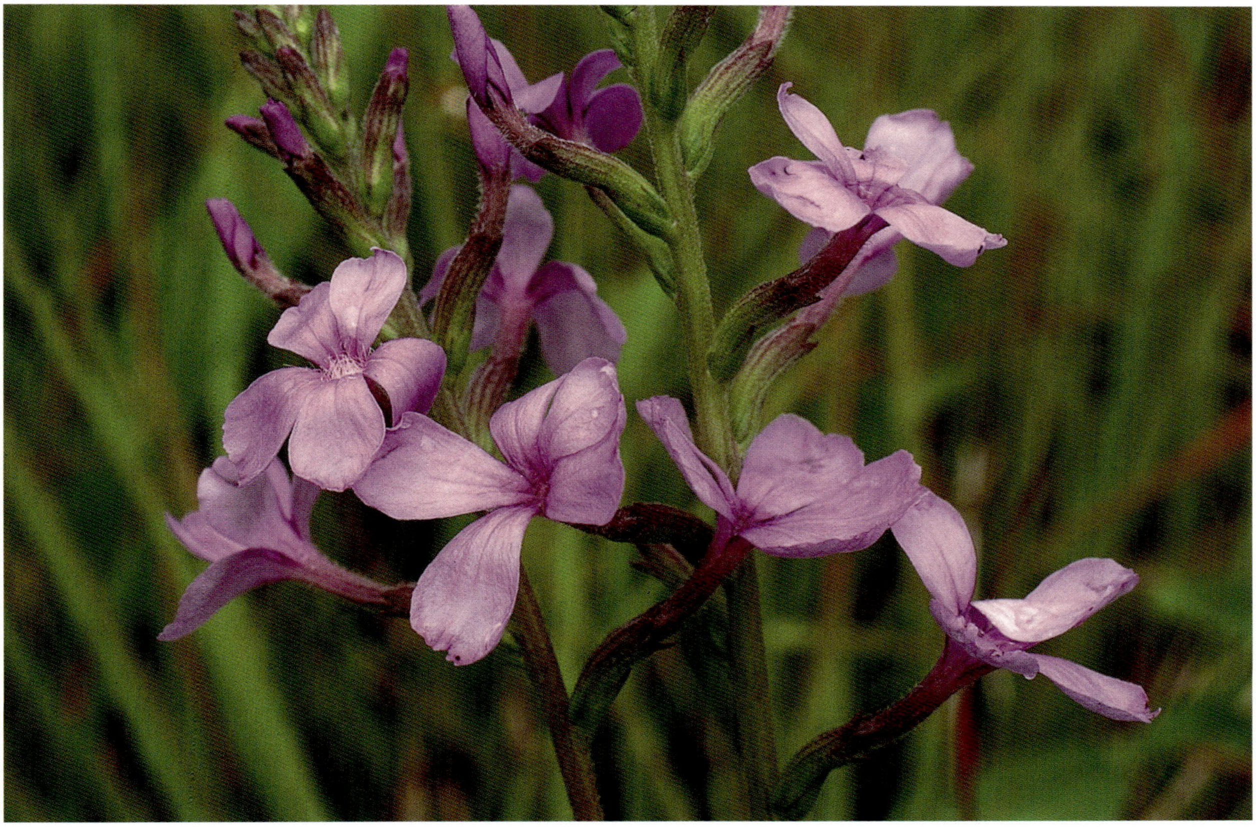

Edge of Appalachia Preserve

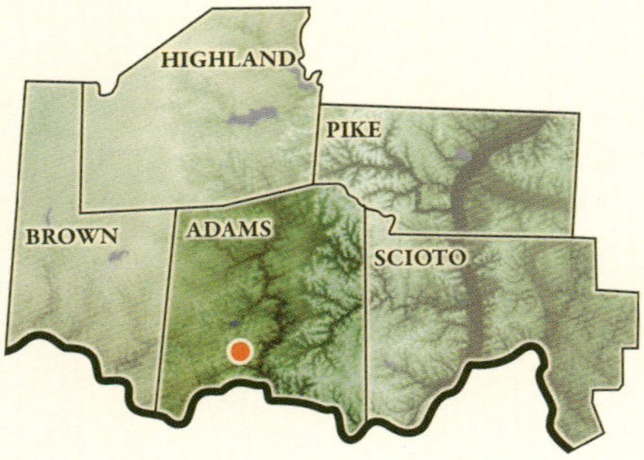

ACREAGE: 13,500

OWNERSHIP: Ohio Chapter of The Nature Conservancy and Cincinnati Museum Center at Union Terminal

ACCESS: Open to the public year-round, dawn to dusk

WEB SITE RESOURCES:
http://www.nature.org/wherewework/northamerica/states/ohio/preserves/art145.html

http://www.cincymuseum.org/explore_our_sites/edge_appalachia/

Centered on Ohio Brush Creek, the sprawling Edge of Appalachia Preserve is the result of a successful collaboration between The Nature Conservancy (TNC) and the Cincinnati Museum Center. Efforts to protect the area, one of Ohio's most biologically significant regions, began in 1959, when the Ohio chapter of TNC purchased its first property, a prairie near the burg of Lynx. Since then, acquisitions have steadily piled up, and the Adams County bioreserve is now one of Ohio's largest protected landscapes.

Geology drives plant communities, and the underlying rock strata are among the oldest in the state, dating to the Ordovician period 450 million years ago. For a hands-on lesson in geology, a hike to Buzzardroost Rock is in order. The Silurian-age limestone that forms this impressive promontory overlooks the scenic Ohio Brush Creek valley. This view from the western boundary of the Allegheny Escarpment is one that all Ohioans should experience.

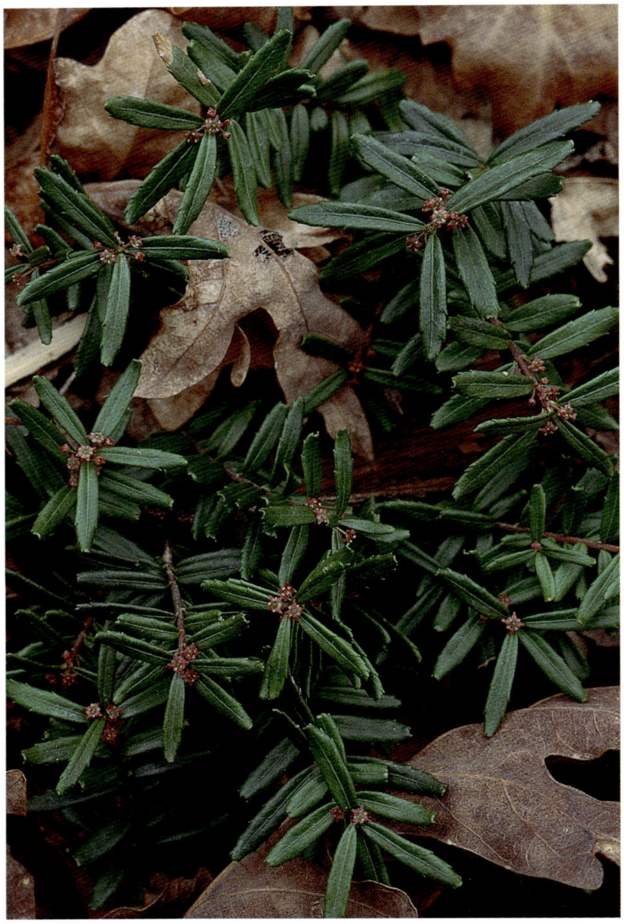

Right: One of our rarest plants, only two small populations of Canby's mountain-lover are known in Ohio. It grows only on dolomite rock. These populations are many thousands of years old, growing vegetatively from an original plant.

Opposite top: An extraordinary view awaits intrepid hikers making the climb to the summit of Buzzardroost Rock. From this dolomite lookout, a seemingly endless expanse of lush forest cloaks the valley of Ohio Brush Creek and the western edge of the Allegheny escarpment.

Opposite bottom: A beautiful, purple-flowered member of the figwort family, ear-leaved foxglove is another in a long list of Adams County megararities. There are only four small populations known to exist in Ohio, all within a few square miles.

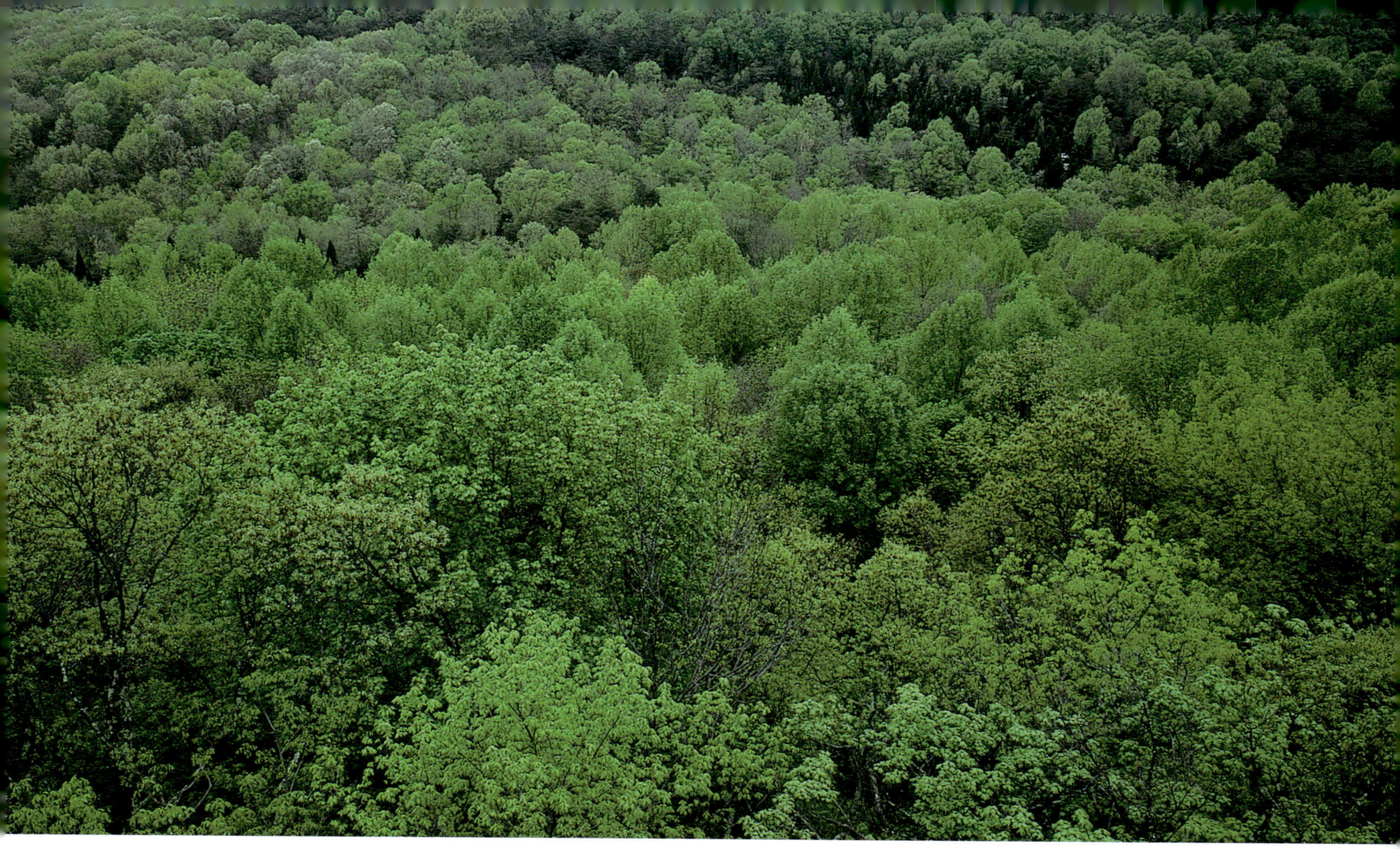

Like the rock below, shortgrass prairies in the vicinity of Lynx are very old—far predating their tallgrass counterparts on the till plains of western Ohio. Botanical treasure troves, these prairies harbor a dazzling array of rarities, including a number of plants found nowhere else in the state. To date, ninety-two species of state-listed plants have been documented in Adams County, one of the densest concentrations of rare flora in the state.

Blazing-stars are jewels of the prairie, their showy purple wands of flowers gleaming in the barrens. Five native species grow in Ohio, and three can be found in prairies of the Edge. Rarest of the lot are stubby slender blazing-stars. They are dwarfed by the magnificent spires of dense blazing-star, which climb two feet or higher. All are butterfly magnets, and their flowers are sometimes covered with scores of nectar-seekers.

A rare and fascinating mammal makes its last stand at the Edge. The Allegheny woodrat, a charming little rodent with bright eyes and a curious disposition, once ranged throughout eastern Ohio and beyond, but has shrunk back alarmingly from northern reaches.

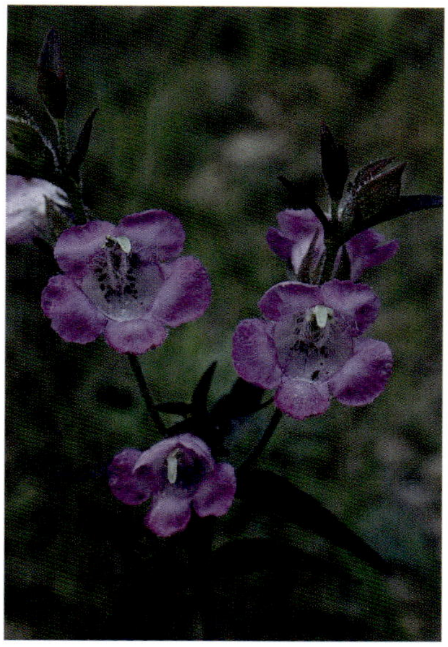

These agile cliff dwellers now occur only in rock formations in Adams County, in Ohio. Consummate packrats, they collect all manner of vegetation, debris, and even man-made objects and pile them into massive trash heaps called middens.

A nighttime visit to the Edge is in order for bird enthusiasts. Warm summer evenings bring the incessantly repeated onomatopoeic song of the chuck-will's-widow. This odd member of the goatsucker family, a close relative of whip-poor-wills, was first discovered in Ohio in 1932 along Ohio Brush Creek. This remains its only stronghold. Their gaping maws are designed for capturing moths, scads of which occur in the Brush Creek valley.

Only a few miles separate the Edge of Appalachia Preserve from one of Ohio's other great ecosystems, the 65,000-acre Shawnee State Forest. Collectively these two sites contain nearly 80,000 acres. Connecting the two would be perhaps the greatest coup in Ohio's short history of conservation.

Dwelling in fissures of dolomite limestone cliffs, green salamanders spend most of their time deep within rocky recesses. They are flattened dorsally for ease of mobility in tight quarters, and their mottled pattern matches lichen-encrusted rocks. Green salamanders emerge at night to forage for small insects along cliff faces.

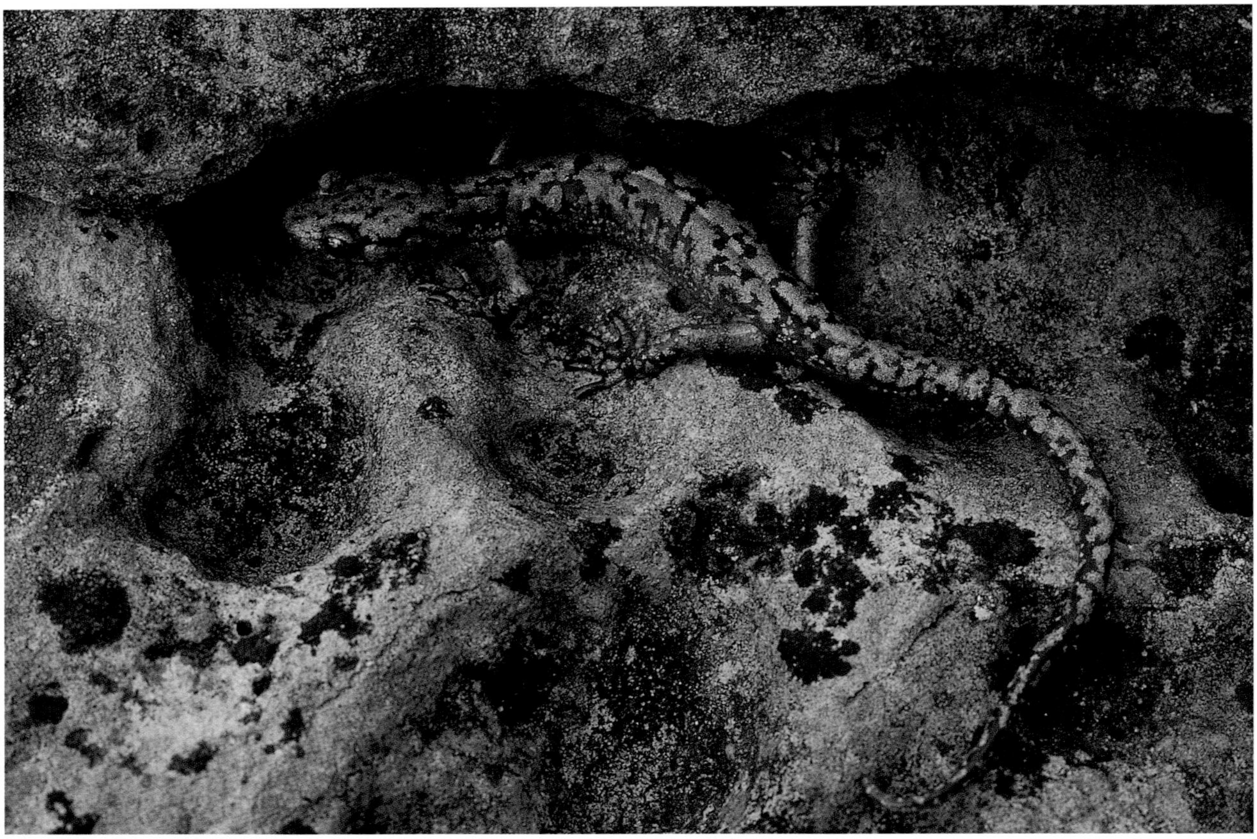

The cedar glade barrens of Adams County are beautiful, exceedingly rare habitats. Grassy balds patchwork red cedar chaparrals interlaced among diverse deciduous forest. A visitor emerging into one of the prairie openings will encounter an astonishing array of flora, including many rare oddities. The last two glaciers never penetrated southern Adams County, and although no one knows exactly when these prairies formed, it was undoubtedly prior to Wisconsin glaciation.

Emma Lucy Braun, a Cincinnatian and pioneer of ecological sciences, put the prairies at Lynx on the map. Early on, she recognized their significance and rare floristic composition. A conservationist long before it became *en vogue,* Braun pushed for the preservation of this singular ecosystem. In 1959, a fledgling group called The Nature Conservancy made fifty-nine acres of Lynx Prairie their inaugural Ohio acquisition. Many important additions have been made since, and Lynx has assimilated into the massive Edge of Appalachia Preserve, one of Ohio's most important natural areas.

Lynx prairies are a botanical showcase, delighting plant enthusiasts from March until October. Leading the parade are tiny white-flowered

Lynx Prairie

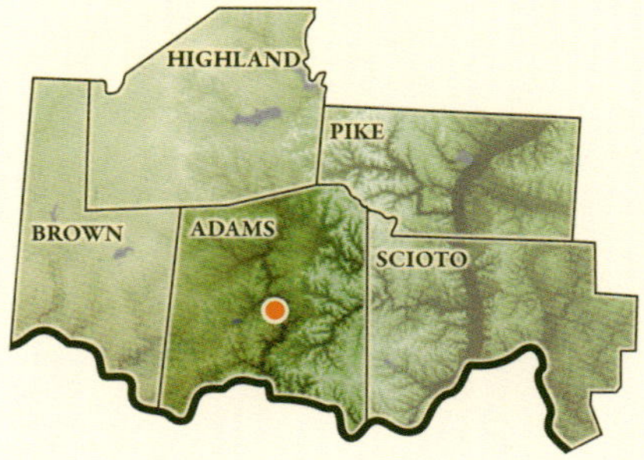

ACREAGE: Approximately 500

OWNERSHIP: Ohio Chapter of The Nature Conservancy and the Cincinnati Museum Center at Union Terminal

ACCESS: Open to the public during posted hours

WEB SITE RESOURCES: http://www.cincymuseum.org/explore_our_sites/edge_appalachia/lynx_prairie.asp

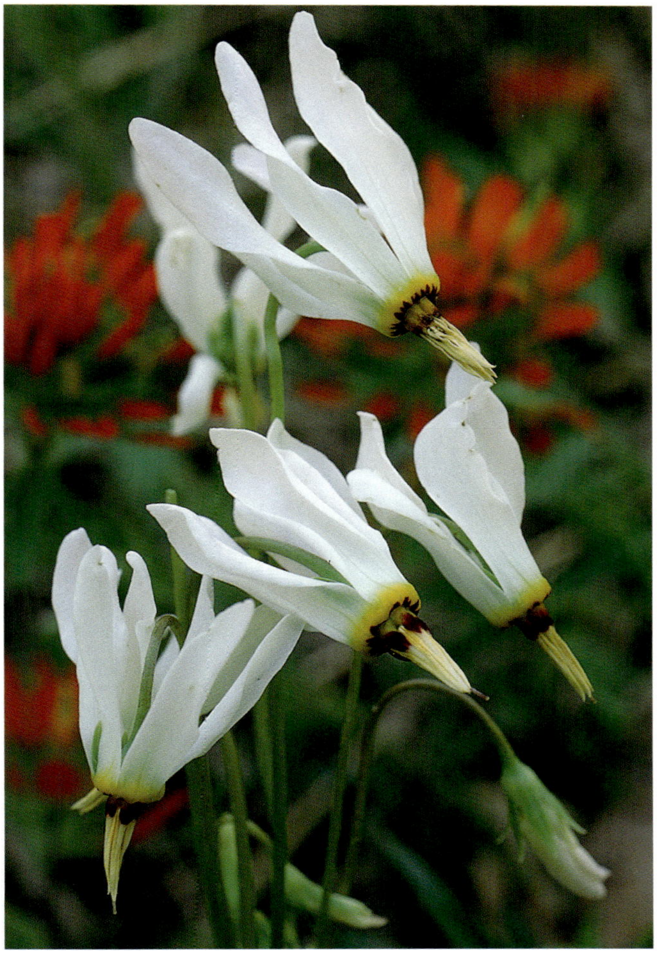

In May, shooting-stars and crimson-orange Indian paintbrush lead the parade of wildflowers that last until the first frosts of autumn.

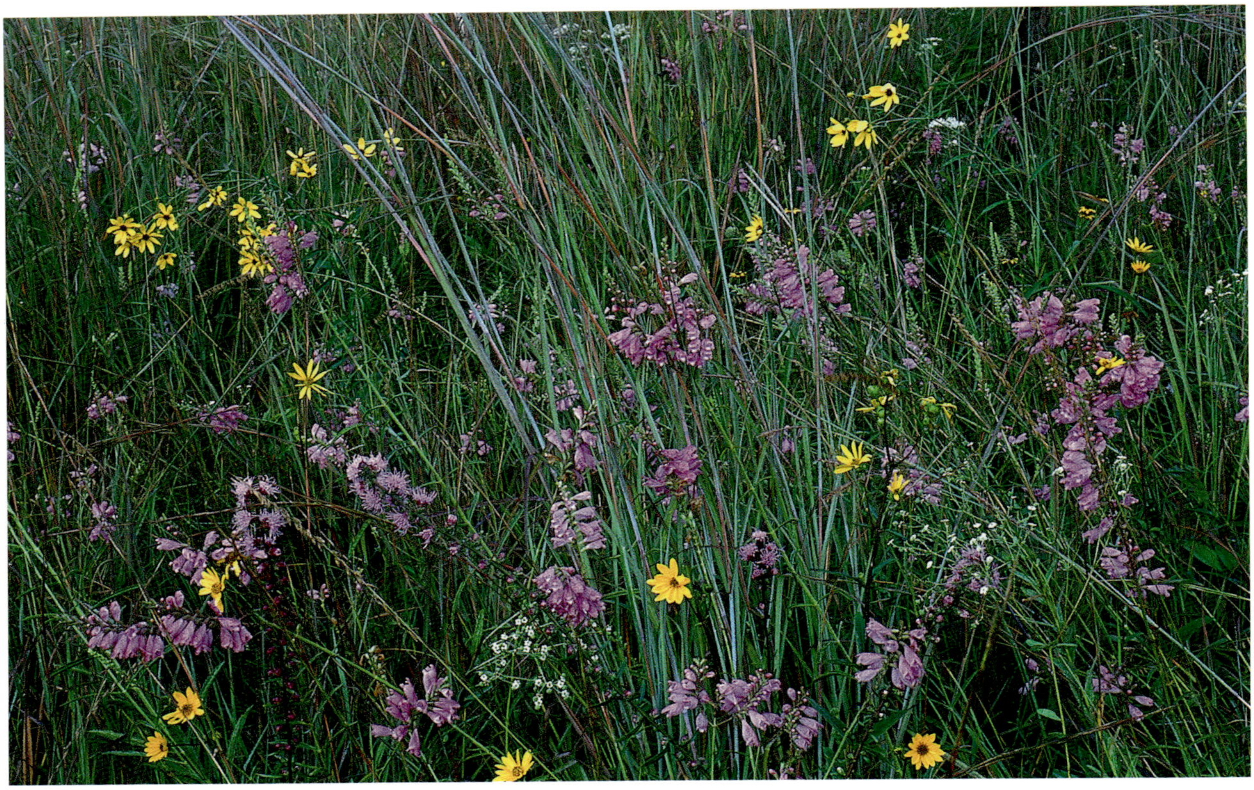

In midsummer, bold purple racemes of obedient-plant grow in profusion amongst grasses in Lynx Prairie. This classic site for Ohio floristics was made famous through the writings of Ohio botanist Emma Lucy Braun.

mustards, wedge-leaved whitlow-grass, and Michaux's leavenworthia. Both are threatened and decorate barren soils of the prairie in March, making them among our first wildflowers to bloom. Scores of flowers follow these pioneers, and each successive week seems to bring new plants as the growing season cascades along.

Nearly forty species of state-listed plants have been documented at Lynx—proof of the botanical importance of these prairies. American aloe long ago moved east from western deserts. Sending forth two-foot naked stalks capped by bizarre greenish blossoms, its fleshy, succulent leaves suggest yucca. Travel to arid western U.S. habitats and you can find several dozen species of aloe, or their closer relatives, agaves.

Summer brings colorful masses of composites: plants in the sunflower family. Scaly blazing-star, Western sunflower, and false boneset shade the prairie in hues of purple, yellow, and creamy white. Those species can't be missed, especially by ace botanists who enjoy finding Lilliputian rarities that lurk in prairie microniches. Perhaps foremost among the obscure is juniper sedge. This globally rare sedge is common in Lynx but wasn't recognized and described to science until 1993.

Meadows teeming with wildflowers are butterfly magnets, and dozens of species occur in the Lynx prairies, including Edward's hairstreak. Gemmed with spots of blue and orange along the hind wings, this quarter-sized beauty is rare and local. Eons of evolutionary gymnastics have allowed it to forge a symbiotic relationship with Allegheny mound ants known as myrmecophily. At night, the defenseless caterpillars enter massive earthen ant nests. Rather than rip the worms asunder, the ants milk them with gentle antenna strokes, prompting secretion of sweet honeydew. In return, phalanxes of guards accompany the caterpillars into the oaks upon which they feed, protecting them from parasitic wasps.

Studies have shown that prairies at Lynx have shrunk by two-thirds since 1938, victims of encroaching forests. One might argue this is natural succession, but perhaps not. People have largely arrested the fires that historically swept such landscapes, keeping the prairies free of overshadowing timber. The Nature Conservancy has reintroduced fire to the Adams County landscape as part of their stewardship, with striking results. Good management should ensure that the botanical treasures of the Bluegrass province flourish for generations to come.

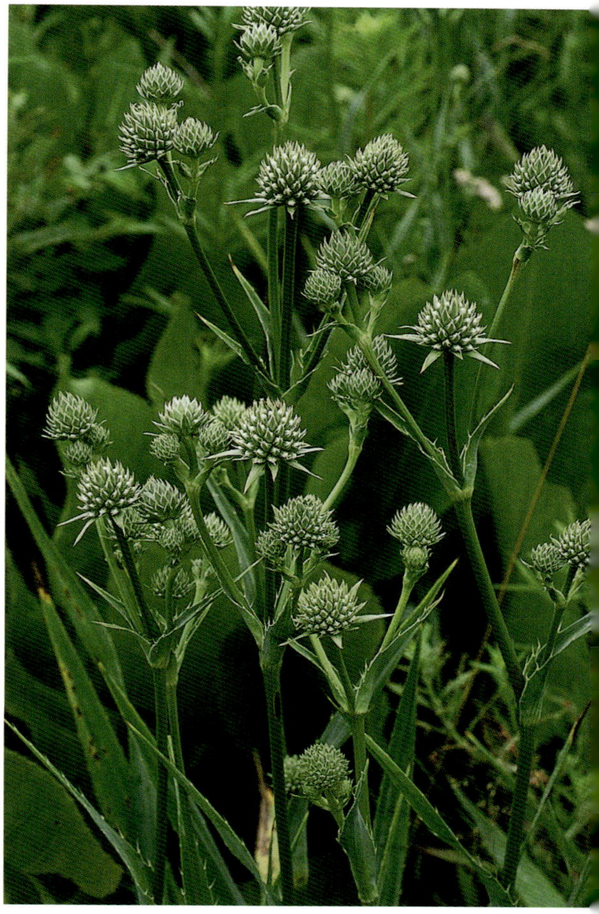

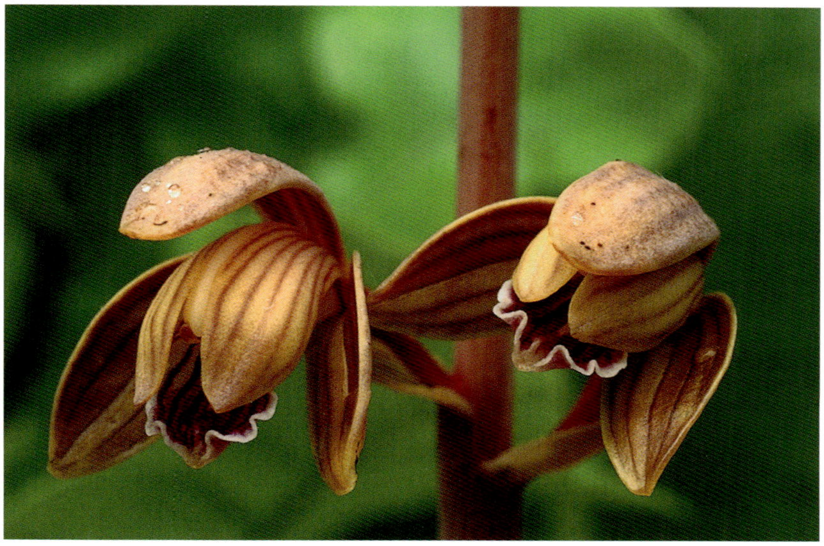

Above: A bizarre-looking member of the parsley family, rattlesnake-master with its curious ball-like heads of greenish-white flowers add an unusual botanical flourish to habitats already rich in strange and rare flora.

Left: A fascinating orchid, crested coral-root is one of many Ohio botanical rarities found only in Adams County. There are five species of coral-root in Ohio, and three, including this one, are endangered.

Central Lowland— Till Plains

Bigelow/Smith Cemeteries State Nature Preserves (10) • Big and Little Darby Creeks (11) • Cedar Bog State Memorial (12) • Clifton Gorge State Nature Preserve (13) • Cranberry Island State Nature Preserve (14) • Killdeer Plains/Big Island Wildlife Areas (15) • Lawrence Woods State Nature Preserve (16) • Mud Lake Bog State Nature Preserve (17)

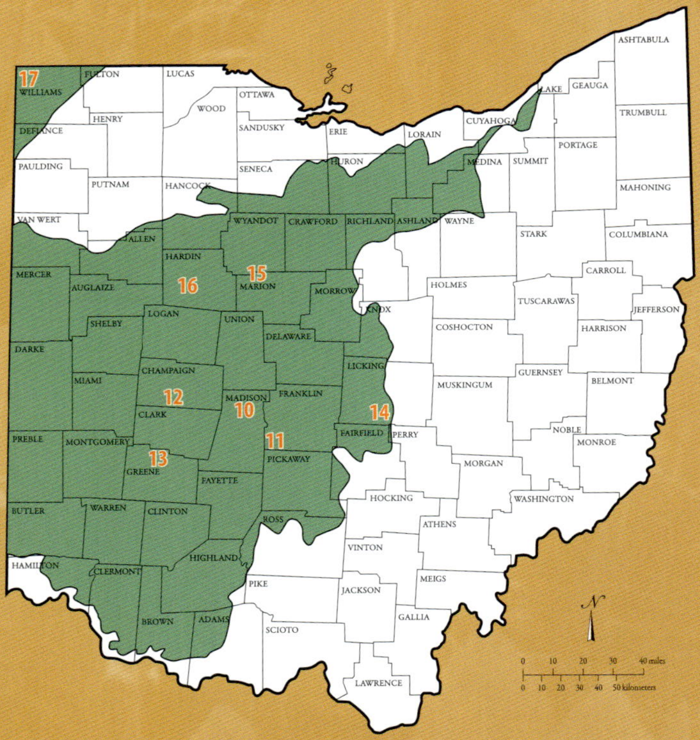

The first pioneers penetrating west-central Ohio ran into a daunting wall of vegetation—the unbroken expanse of tallgrass prairie of the Darby Plains. Sprawling over some 385 square miles, the prairie was dominated by Indian grass and big bluestem. These grasses grew so tall that a man on horseback could scarcely see over them.

Fire and prairie go together, and the Darby Plains regularly went ablaze. These conflagrations must have been awe inspiring but horrific spectacles to settlers. Dr. Jeremiah Converse, an early pioneer, wrote: "The blaze of the burning grass seemed to reach the very clouds . . . [flames] would leap forty or fifty feet in advance of the base of the fire. Then add to all this a line of the devouring element three miles in length, mounting upward and leaping madly forward with lapping tongue, as if it were trying to devour the very earth, and you have a faint idea of some of the scenes that were witnessed by the early settlers of this country" (Jeremiah Converse, History of Darby Township, Ohio, 1883).

As fearsome as fire can be, it is essential to prairie ecosystems. Prairie plants co-evolved with fire and are adapted to the specialized conditions that fire causes. Burning doesn't affect deep-rooted prairie forbs and grasses but kills off shallowly rooted non-native grasses and invading woody plants.

Today, nearly nothing remains of the original Darby Plains prairie. This is true of Ohio's prairies in general; over 99 percent have given way to the plow or development. That's why these postage-stamp-sized graveyards are so special. As they were founded before the area

Bigelow/Smith Cemeteries State Nature Preserves

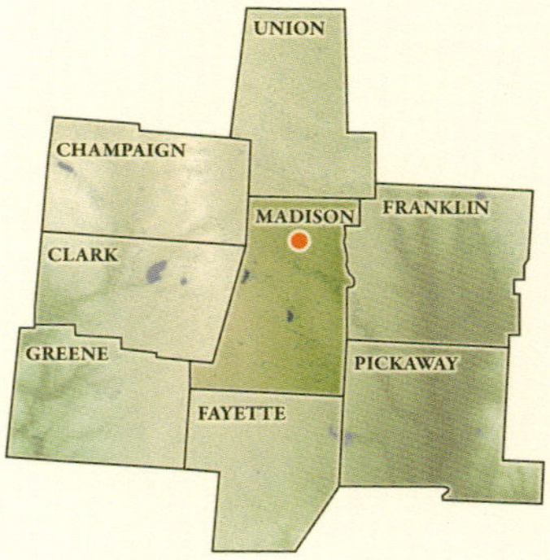

ACREAGE: 1½ (Bigelow ½; Smith 1)

OWNERSHIP: Pike and Darby townships (Bigelow and Smith, respectively) of Madison County. Managed by the Ohio Division of Natural Areas and Preserves.

ACCESS: Open to the public during posted hours

WEB SITE RESOURCES:
http://www.ohiodnr.com/dnap/location/bigelow_prairie.html

http://www.ohiodnr.com/dnap/location/smith_cemetery.html

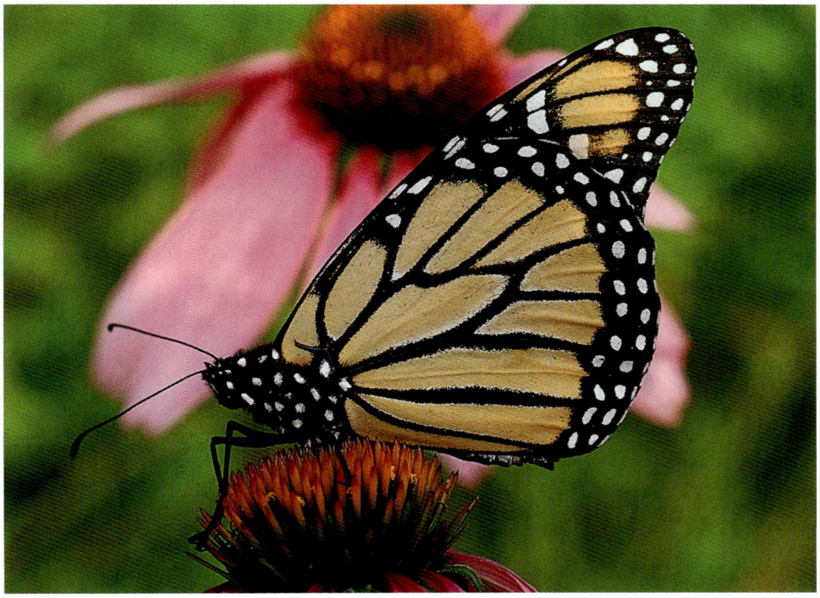

One of the most amazing stories of the insect world is the annual peregrination of Monarchs to wintering grounds in lofty Mexican fir forests. From Ohio and points north, Monarchs wing their way south. Prairie remnants often host large numbers, as the butterflies gather to feed on blooms of various wildflowers.

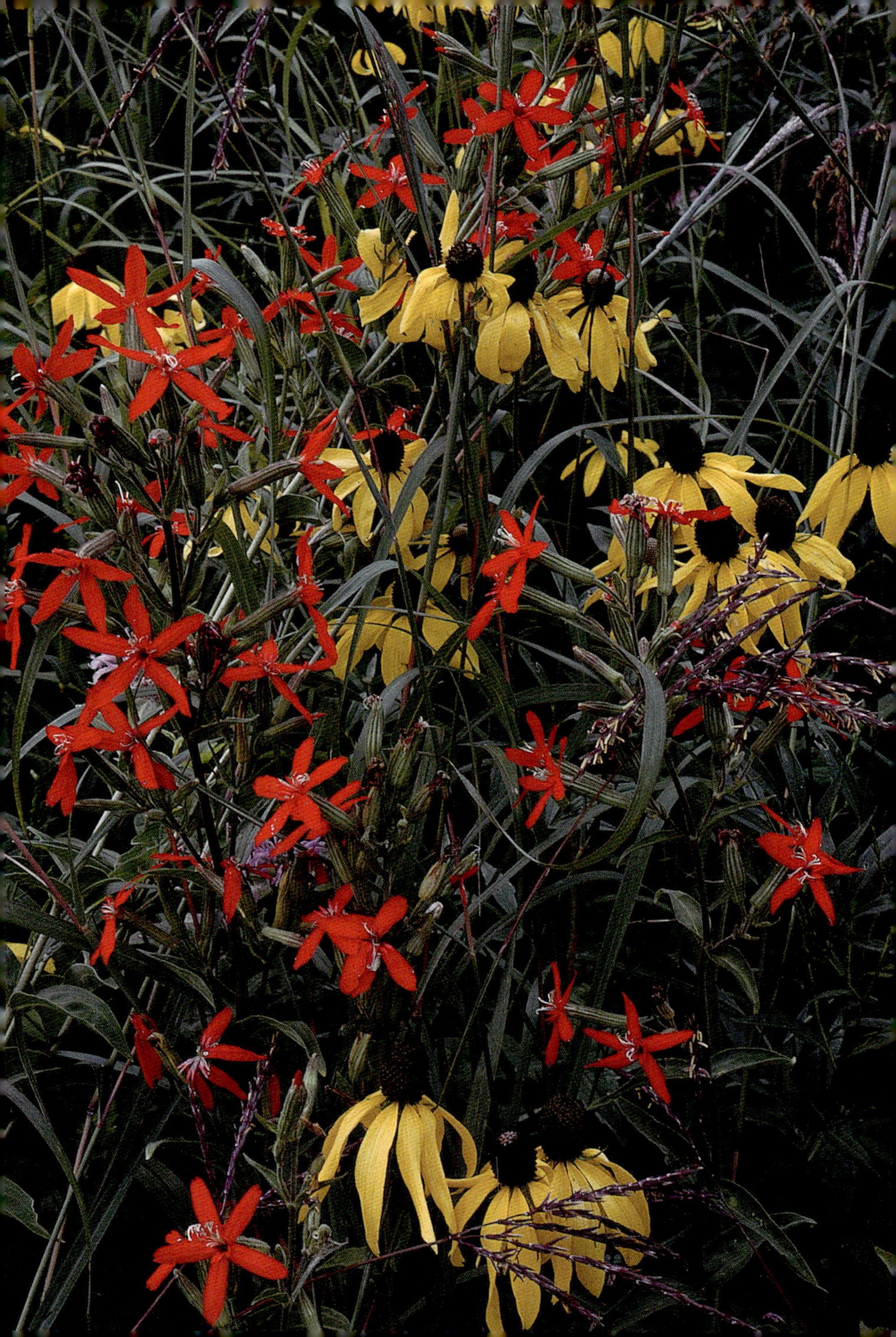

was settled and tamed, Bigelow and Smith cemeteries harbor some of the only unplowed prairie sod in Ohio. Pioneers probably augmented the cemeteries with showy flowers as grave decorations, brought from surrounding prairie. Bigelow and Smith are the prairie equivalent to virgin forest, and the flora of the cemeteries offers a tiny glimpse into the spectacular, primeval prairie that our first settlers encountered.

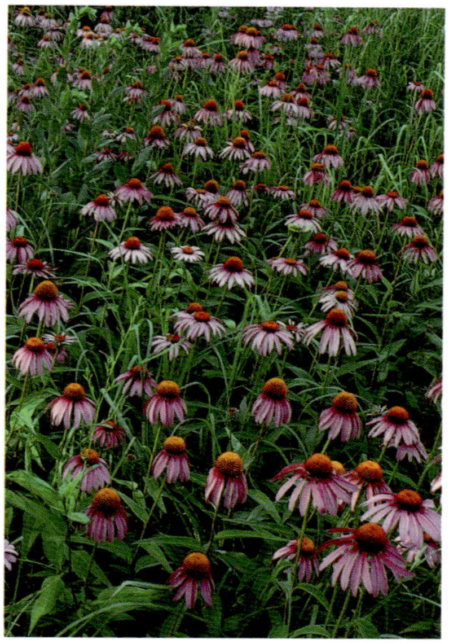

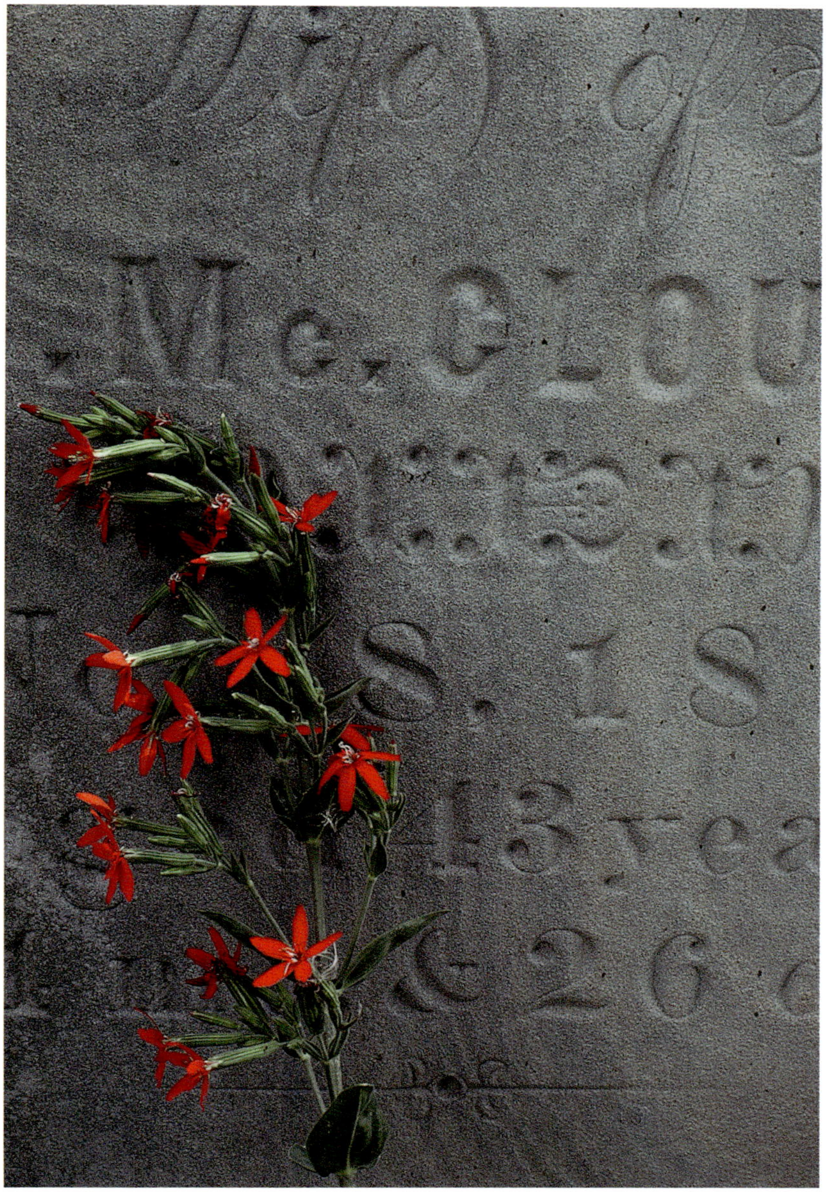

Opposite: Few plants are as characteristic of Ohio's former prairie regions as royal catchfly. It was probably abundant in the presettlement prairie landscape. Today it is considered threatened and found in few locales. The odd name "catchfly" refers to the sticky calyx—the cup subtending the flower.

Left: Midsummer flowers bloom in profusion in small but spectacular Bigelow Cemetery. Sprays of scarlet royal catchfly and gray-headed coneflower brighten the faded headstones of prairie pioneers.

Above: In the wild, showy purple coneflowers are nearly always associated with prairie remnants. Lush stands cloak Smith Cemetery in a blanket of midsummer purple.

Big and Little Darby Creeks

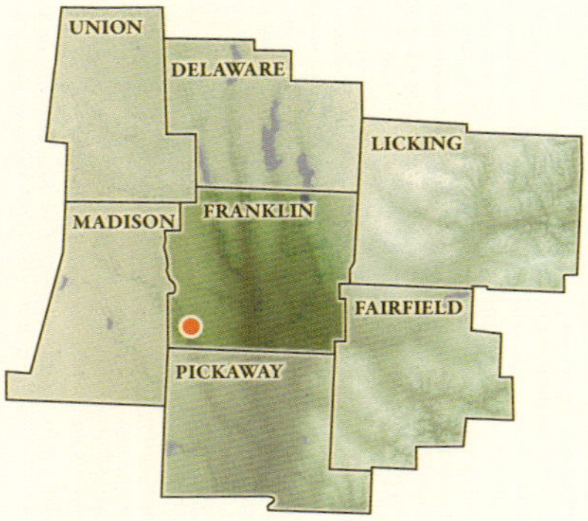

Circle denotes location of Battelle Darby Creek Metro Park, one of the best sites to access the streams.

ACREAGE: Combined properties of Franklin County Metroparks, TNC, and ODNR total over 7,000 acres

OWNERSHIP: Miscellaneous

ACCESS: Best access via Columbus and Franklin County Metroparks. Metroparks open to the public during posted hours.

WEB SITE RESOURCES:
http://www.metroparks.net/

http://www.ohiodnr.com/dnap/sr/bdarby.htm

http://www.nature.org/wherewework/northamerica/states/ohio/bigdarby/

Flowing eighty-eight miles south from headwaters in the rolling hills of Logan County, the Darby confluences with the mighty Scioto River at Circleville. In spite of close proximity to the major metropolis of Columbus, this stream remains one of the most pristine in the country.

The Nature Conservancy declared the Darby watershed one of the "Last Great Places" in North America. Rightly so; this national treasure harbors an incredible array of aquatic life. Of 166 species of fish documented in Ohio, 103 are known from the Darby and its tributaries, including one that is likely extinct and was Ohio's only endemic fish. On November 4, 1943, famed ichthyologist Milton B. Trautman was seining a fast-flowing riffle in Pickaway County when he netted a small, unknown catfish. It turned out to be new to science and was named the Scioto madtom. Collected on eight separate occasions afterward, it was last caught on November 17, 1957. It hasn't been seen since.

Yet life aplenty remains in Darby waters, including a spectacular group of fishes—the warblers of the underwater world. Darters, tiny members of the perch family, live primarily in turbulent waters of riffles and chutes, foraging among gravelly substrates. Males morph into gaudy hues in late winter, rivaling anything found in domestic aquaria.

Hardest hit of all aquatic animals in Ohio are freshwater mussels, or unionids. Because of their specific foraging tactics, burrowing into stream bottoms and filtering out small organisms, mussels are vulnerable to smothering by sediment runoff. Fourteen of the eighty species

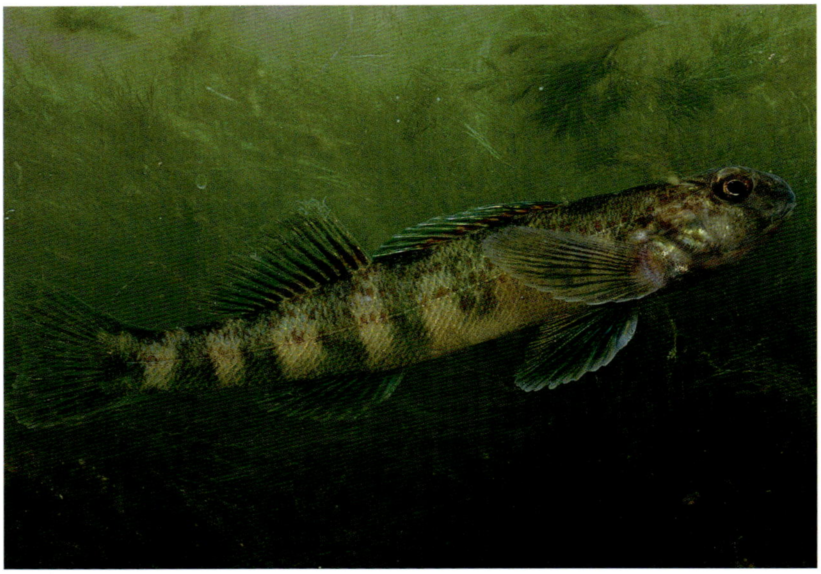

One of twenty-four species of darters recorded in Ohio, greenside darters are common in the Darby and throughout much of Ohio. Males in breeding colors are striking. Unlike most fish, darters lack air bladders. This adaptation allows them to easily anchor on the stream bottom in the fast-flowing torrents that they favor.

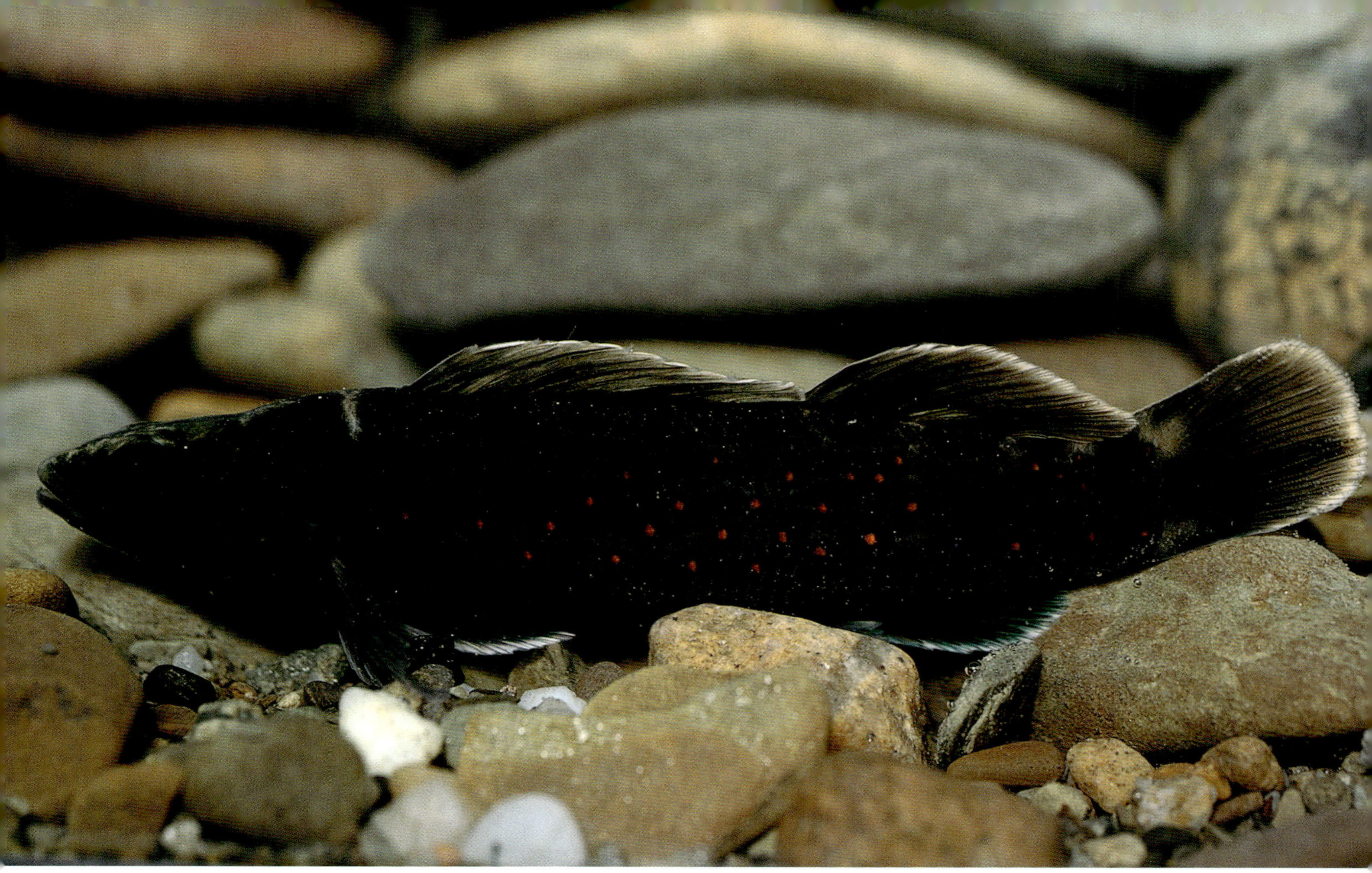

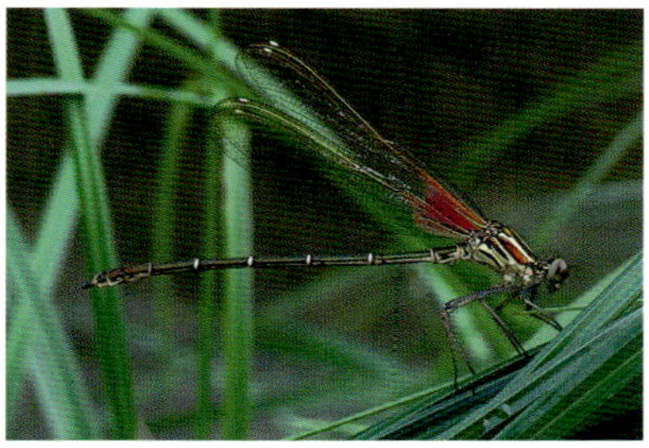

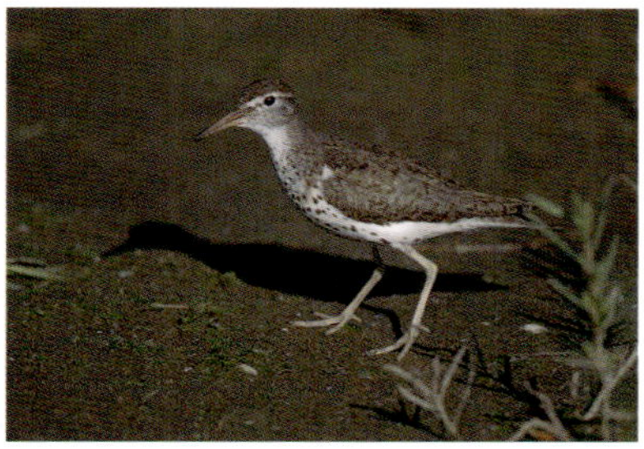

Above: The best populations of endangered spotted darters occur in the Darby, which is a stronghold for other rare fish and aquatic organisms.

Far left: Beautiful American rubyspots are common along the edges of Ohio streams, particularly around beds of water-willow.

Left: Of the forty-seven species of sandpipers and plovers reported in Ohio, six remain to nest, and only three are common breeders. The spotted sandpiper nests along many Ohio waterways. A characteristic behavior is their habit of bobbing their tail.

once known to occur are extirpated from Ohio. The Darby, because of its excellent water quality, still harbors at least thirty-eight mussel species, including eight listed as endangered. Two of them, Northern riffleshell and clubshell, are federally endangered—the rarest of the rare.

Conserving streams and associated riparian corridors may have greater largescale impacts that any other preservation efforts. Protecting the Darby sends purer water into the Scioto River and then to the Ohio River. That great stream, which drains much of the Midwest, flows into the mighty Mississippi, then into the Gulf of Mexico. There is now a dead, deoxygenated zone the size of New Jersey in the gulf at

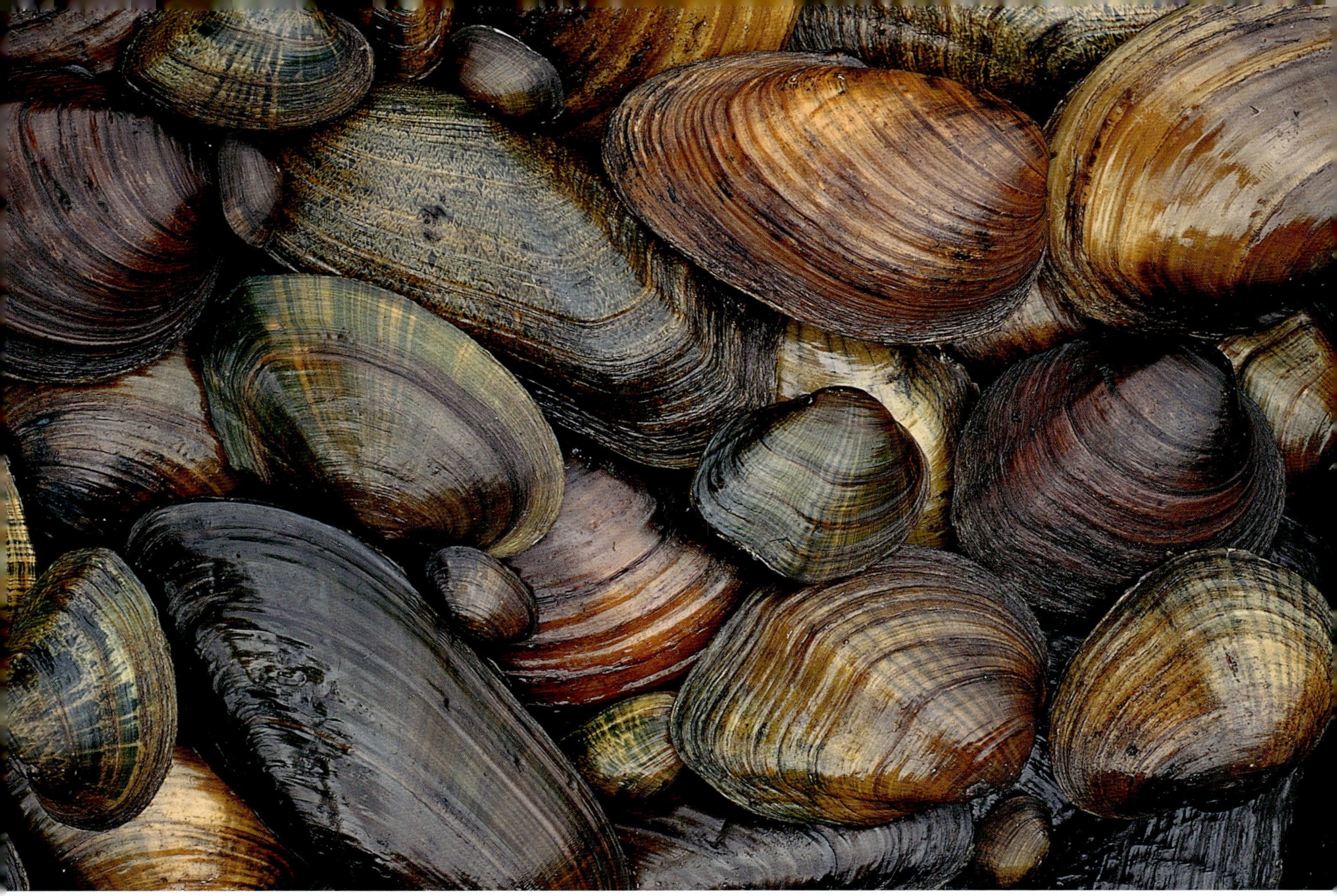

Although seemingly dull, freshwater mussels have many interesting names. Pictured above are at least ten species. They include pink heelsplitter, lilliput, rabbitsfoot, and clubshell. Mussels have a reproductive cycle that defies belief. Some species lure suitable host fish near with an enlarged mantle—fleshy appendages that look like tasty food. When the fish draws near, the mussel blasts it with a spray of glochidia, or larval clams. The glochidia mature within the gills of the host fish and, once of a suitable size, drop off into the stream bottom. There, they may live for many decades.

the Mississippi's mouth. This aquatic desert is caused mainly by farmland fertilizers and sewage running into the vast labyrinth of streams that form the Mississippi drainage.

Big Darby Creek and its major tributary, Little Darby, were designated a State Scenic River in 1984. A further recognition of wildness was awarded in 1994, when it received National Scenic River status by the federal government. As Ohio's capital city, Columbus, sprawls outward, protecting the Darby will become one of our greatest environmental challenges. Hopefully we are up to the task.

Stepping into Cedar Bog is akin to taking a leap back into the Pleistocene age. Many plants and animals here are glacial relicts, holdovers from long ago. A visitor will instantly notice something unique at this site. No other Ohio wetland is cloaked with dense growths of white cedar. This northern conifer is the defining feature of Cedar Bog, but only one of its many rare plants.

While the name is charismatic, Cedar "Bog" is actually a fen. "Bog" is an oft-employed catchall phrase for various wetlands. True bogs are glacial kettle lakes that fill with water and, over time, distinct vegetation. Fens flush. There is a constant energetic flow of water derived from subterranean aquifers. By the time cold artesian water bursts to the surface, it is supercharged with calcium precipitates derived from long passages through limestone beds. Fens are essentially big sheets of water upon which rests a blanket of highly specialized plants. The combination of high pH, cold root-level temperatures, and saturated soils creates growing conditions too tough for most plants.

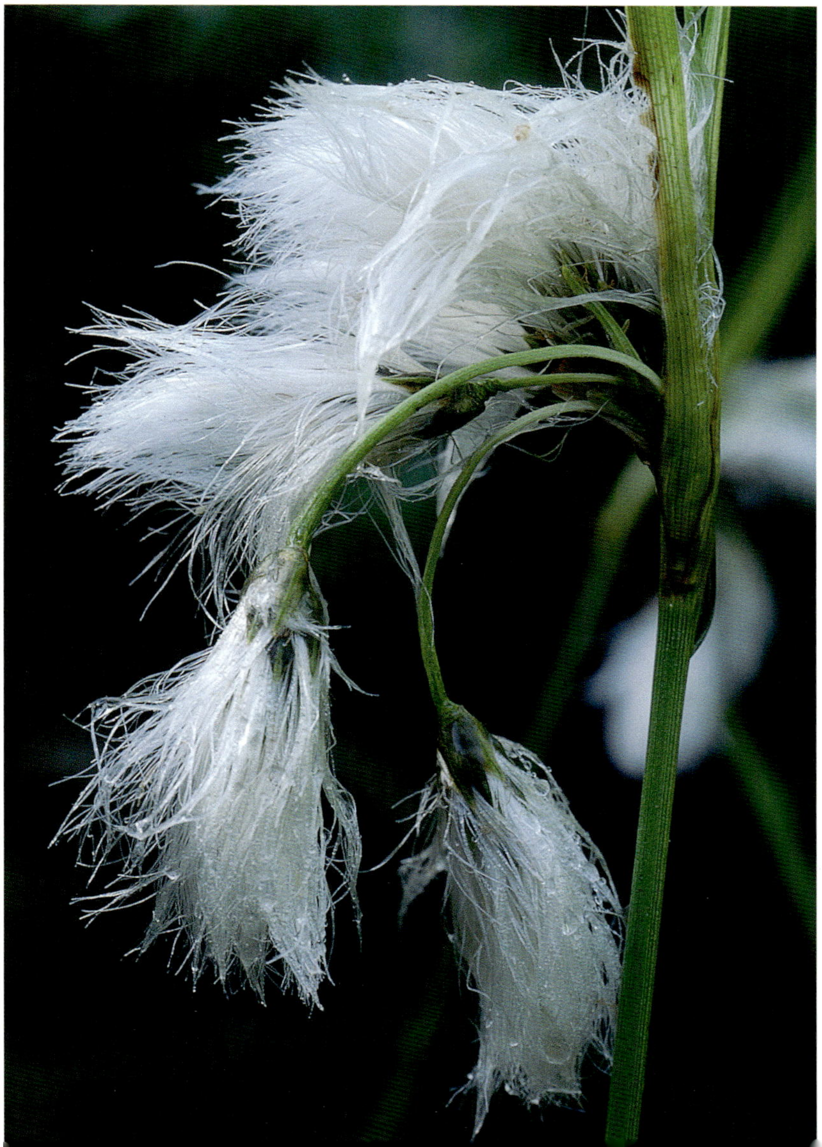

Cedar Bog State Memorial

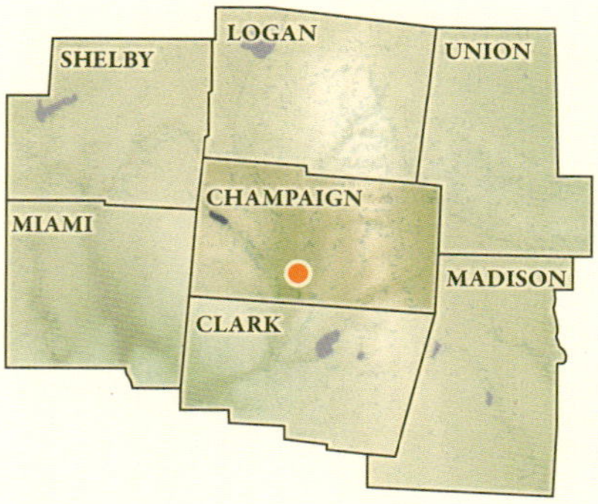

ACREAGE: 427

OWNERSHIP: Ohio Historical Society

ACCESS: Located at 980 Woodburn Road, one mile west of U.S. Route 68 and about two miles south of Urbana. Open during posted hours. Donations are very welcome at entrance kiosk.

WEB SITE RESOURCES:
http://www.cedarbog.org/index.htm

http://www.ohiohistory.org/

http://www.ohiodnr.com/dnap/location/cedar_bog.html

Fens are built on sedges. Most plant biomass in Cedar Bog's meadows is sedges, and as they die and decompose, gramineus peat is formed. Green cotton-grass (actually a sedge) is an Ohio rarity, as most cotton-grasses are true northerners. Only two species occur in Ohio, and both are scarce.

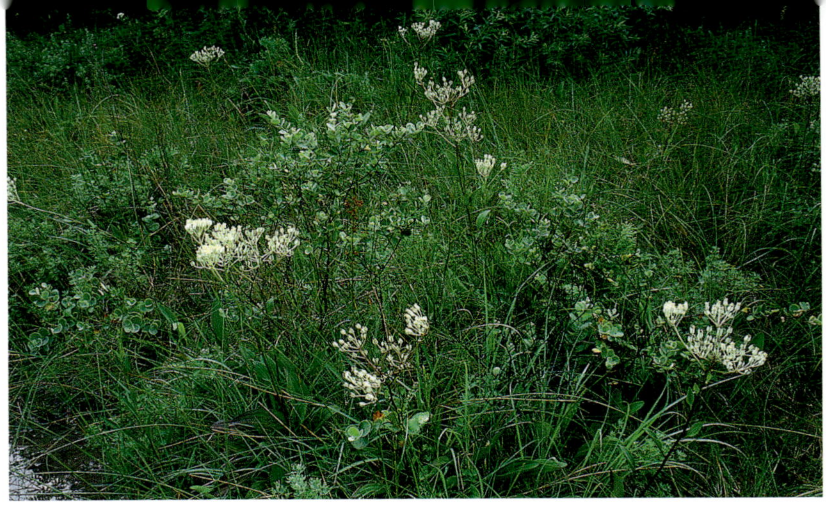

Right: Creamy blossoms of the rare fen Indian-plantain ornament Cedar Bog meadows in June. Cedar Bog is filled with rare plants that are extremely habitat specific and often typical of boreal habitats found far to the north.

Below: Endangered in Ohio, seepage dancers are known from only four other sites. Cedar Bog hosts the largest population of this tiny damselfly.

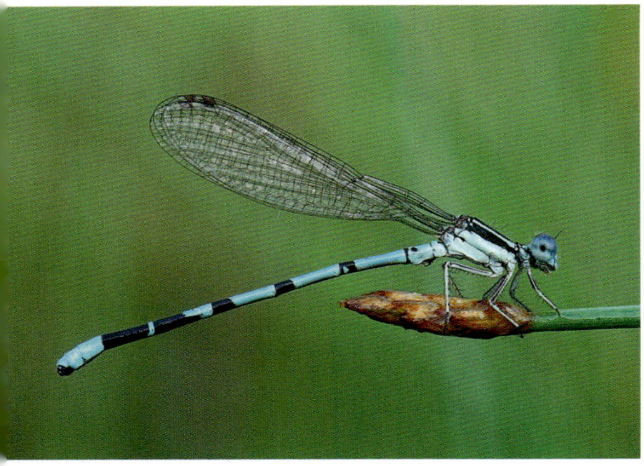

Rare plants abound here, foot for foot, more than any other Ohio habitat. At least thirty-five state-listed rarities occur, including some of the state's most endangered species. One of the most spectacular is North America's largest orchid, the showy lady's-slipper. Pink and white blooms are the size of a small fist, and an opening in the pouch-shaped flowers accommodates large bumblebees, their principal pollinator. Other noteworthy plants include bog aster, prairie valerian, shrubby cinquefoil, and wand-lily.

A few species of carnivorous plants adorn Cedar Bog's meadows. These miniature wonders of evolution have amazing strategies of coping with nutrient-deficient fen soils. The endangered horned bladderwort grows in the most alkaline barrens of marl flats, pushing forth beautiful yellow violetlike flowers. Their slender roots are beset with tiny bladders that capture miniscule animals. Bladders are armed with guard miniscule hairs surrounding a trap door of sorts, and prey is lured by specialized cells that secrete sugary chemicals. If a victim trips a few guard hairs, the trap opens inward so forcefully that the animal is instantly sucked inside. The door slams tight and internal secretions break down the bug, making its nitrogen and proteins available to the plant.

Rare animals persist at Cedar Bog, and two of the most interesting are the Eastern Massasauga rattlesnake and the spotted turtle. Both of these species have become scarce in Ohio as their wet prairie habitat has been largely destroyed. The turtle is our smallest species; a big one is five inches long. Shiny black and polka-dotted with yellow, spotted turtles are secretive and lurk in wet sedge meadows. Furtive in the extreme, Eastern Massasaugas are hard to find and sometimes live in crayfish burrows. They average about two feet in length and are often docile, but like any poisonous snake they can bite quickly. A colloquial name is black snapper (there are also many other colloquial names).

The 427 acres of Cedar Bog is a shard of the former seven-thousand-acre fen that once covered this region. Protecting what's left is not without challenges. In the 1960s plans were forged that would have routed U.S. Route 68 hard on the east border of Cedar Bog. That intrusion would have been detrimental to the fen's delicate ecology, and environmentalists successfully derailed the scheme. Today's biggest challenge may be protection of groundwater that feeds the wetland. As development expands locally, degradation and disruption of hydrology become more likely and safeguarding wide-ranging water sources won't be easy.

Few Cedar Bog visitors will be fortunate enough to glimpse the Eastern Massasauga. It is rare anywhere in Ohio. Most of its original wet prairie habitat has been converted to cropland.

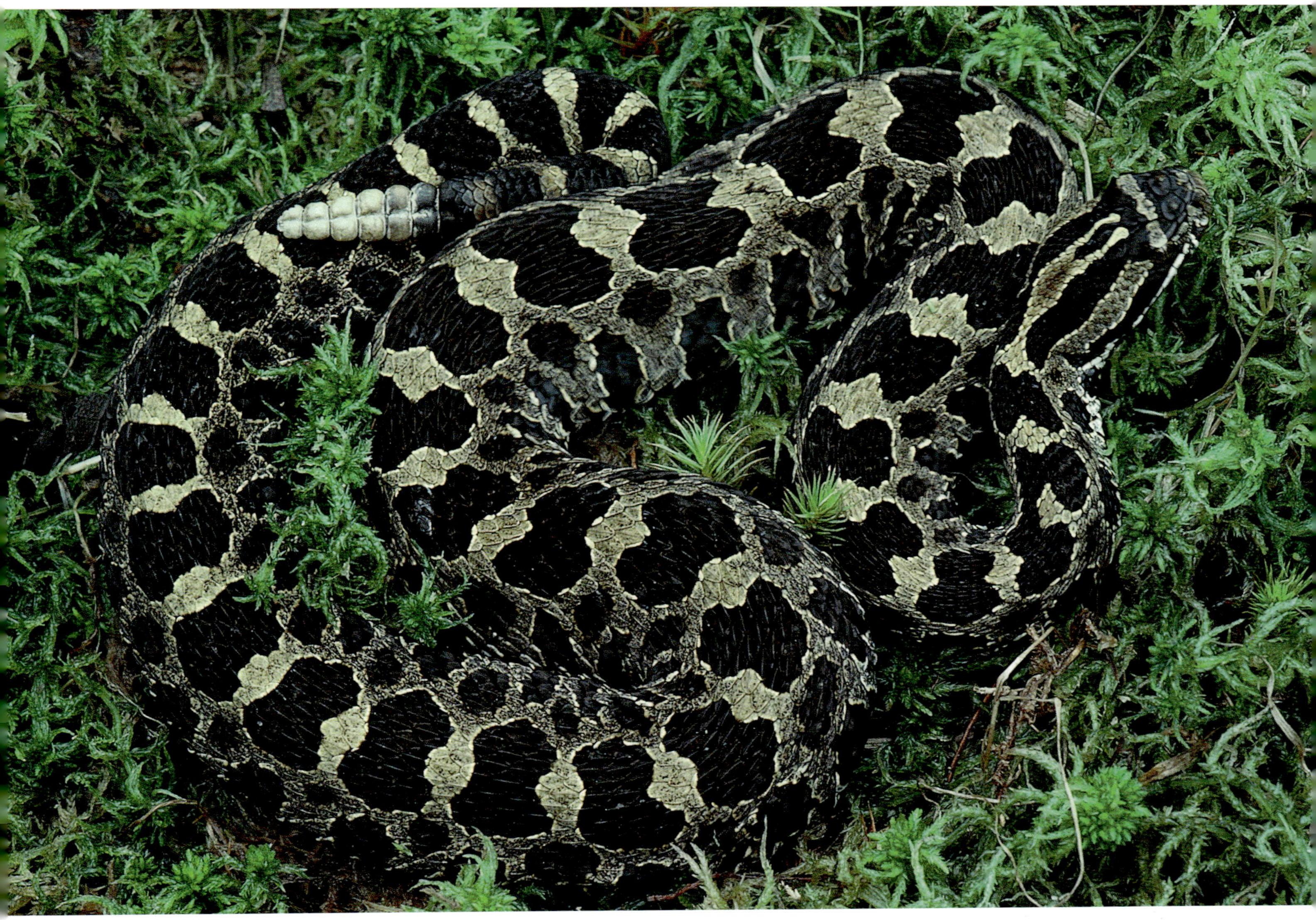

Clifton Gorge State Nature Preserve

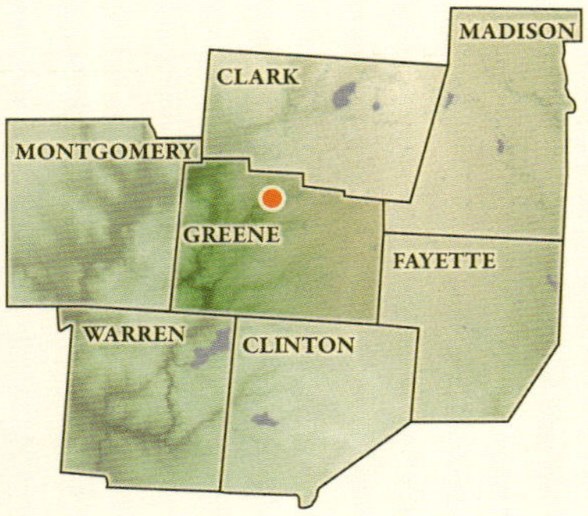

ACREAGE: 269

OWNERSHIP: Ohio Division of Natural Areas and Preserves

ACCESS: Open during posted hours

WEB SITE RESOURCES:
http://www.ohiodnr.com/dnap/location/clifton.html

http://www.ohiodnr.com/parks/parks/jhnbryan.htm

http://www.ohiobirds.org/birdingsites/showsite.php?Site_ID=5

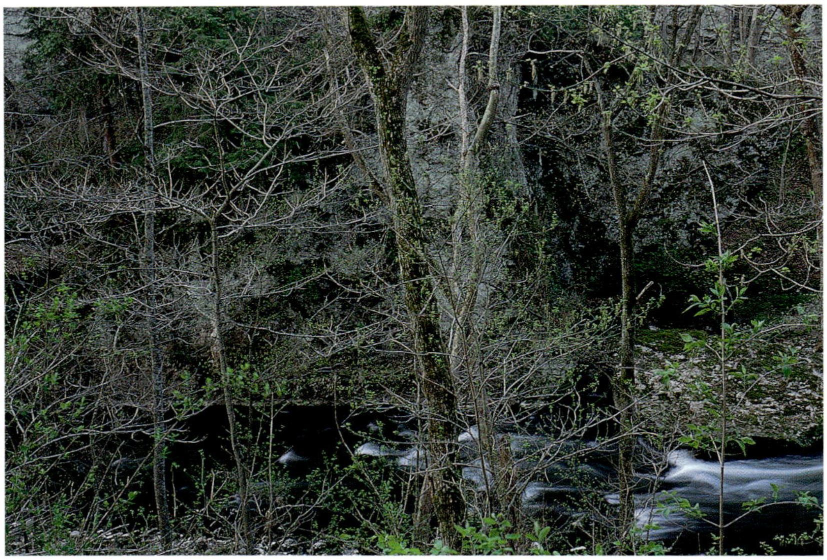

The waters of the Little Miami River sluice through the rocky confines of Clifton Gorge. Steep walls of sheer limestone contain the stream, which jigs around gigantic slump blocks calved from the cliffs long ago. Ferns and other interesting flora grow thickly on the bedrock, and rare white cedars shoot precariously from rock-face crevasses.

You will often hear Clifton Gorge before you see it. The swift, rushing waters of the Little Miami River create a low roar that drifts up to the rim trails, providing an aural greeting for hikers.

This is an outstanding example of interglacial and postglacial stream cutting. Clifton Gorge's birth dates to the Silurian age, over 400 million years ago. Then, a large sea covered what is now the Buckeye State. As this sea gradually receded, dolomite limestone—the substrate of Clifton Gorge—was formed by crystallization of various minerals. Much later, as modern patterns of drainage evolved, water exploited weaknesses in the rock and began carving the vast fissure today's visitor sees.

Sheer rock walls line either side of the river, creating an impressive canyon unmatched elsewhere in western Ohio. Because of the depth of the gorge and the streams of subterranean fifty-six-degree air jetting from crevasses, the depths of the valley stay cool and shaded. This specialized microclimate allows northern plants to abound—far south of most of their brethren.

Boldest of the northern vegetation are white cedars, rooted precariously on the limestone escarpments and slump blocks. Plenty of botanical rarities can be found, most of which rarely occur this far south. A practiced eye might spot the crimson fruit of red baneberry, snowy blossoms of rock serviceberry, or the showy false melic grass. In all, over a dozen species of state-listed rare plants have been found.

The vertical cliffs and underlying talus slopes are a fern lover's paradise. A few dozen species can be found, including cliff-face oddities such as smooth cliff-brake and mountain spleenwort. Streamside

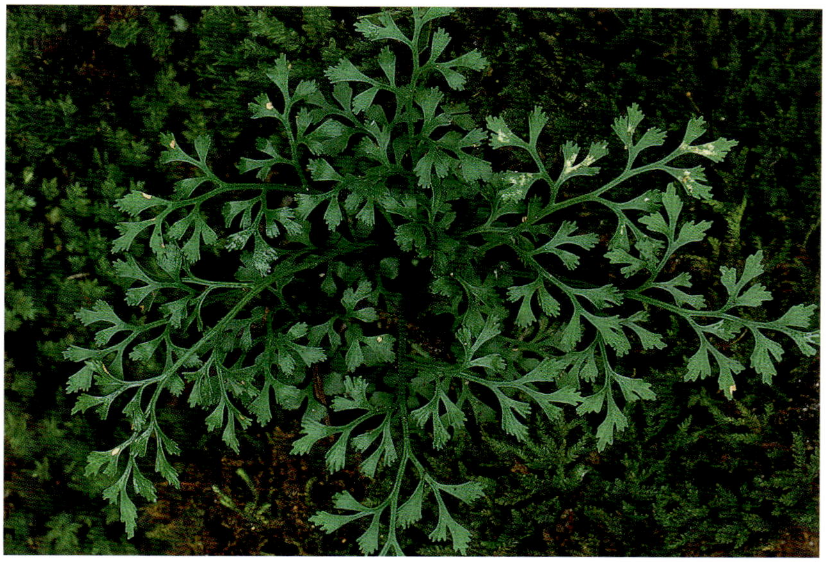

Left: A diminutive rarity, the threatened wall-rue is found in but a few Ohio sites. It is one of about twenty of Ohio's eighty-five species of ferns that is an obligate saxicole—it only occurs on cliff faces.

Below: Ohio plays an important role in the history of snow trillium. It was first discovered by pioneer botanist John Riddell in 1834 along limestone cliffs bordering the Scioto River in Franklin County. It also occurs in Clinton Gorge.

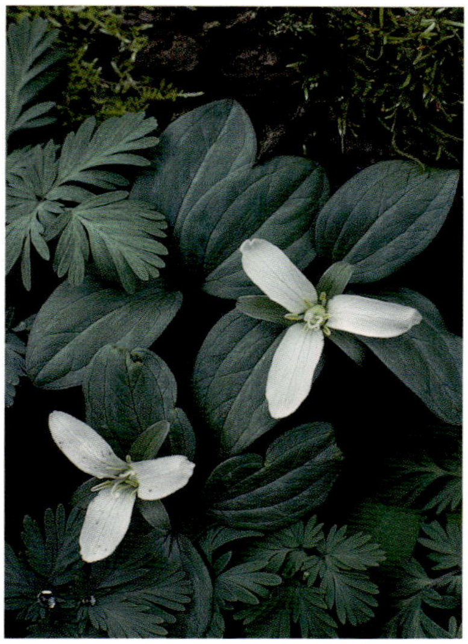

boulders are covered with colonies of walking fern, an interesting plant that arcs its leaf tips outward until they touch rock, whence they grow roots and shoot out a new leaf. Over time, they walk their way into new terrain and enlarge the colonies.

Centerpiece of the gorge is the Little Miami River, among our most pristine streams. The first river to be dedicated as scenic in Ohio (1969), it has also been federally designated as a scenic river, the first waterway to receive that status. Nearly ninety species of fishes and three dozen freshwater mussel species live in its waters—exceptional diversity and proof of the Little Miami's outstanding aquatic habitats.

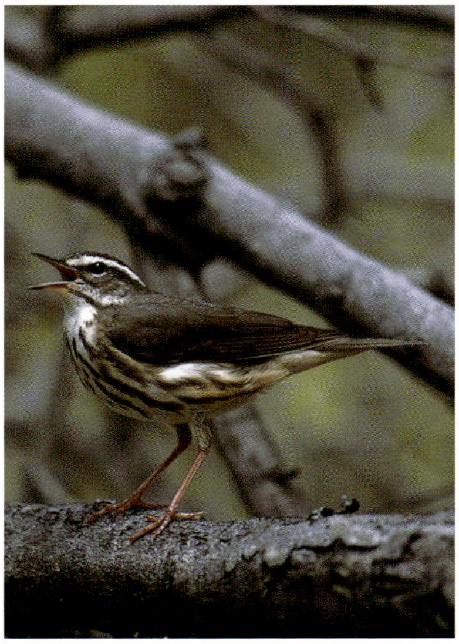

Louisiana waterthrushes are a true harbinger of spring. One of about sixty species of neotropical birds that nest in Ohio, the waterthrush is one of the first warblers of this group to return. Their loud rollicking song fills Clifton Gorge beginning in late March.

Cranberry Island State Nature Preserve

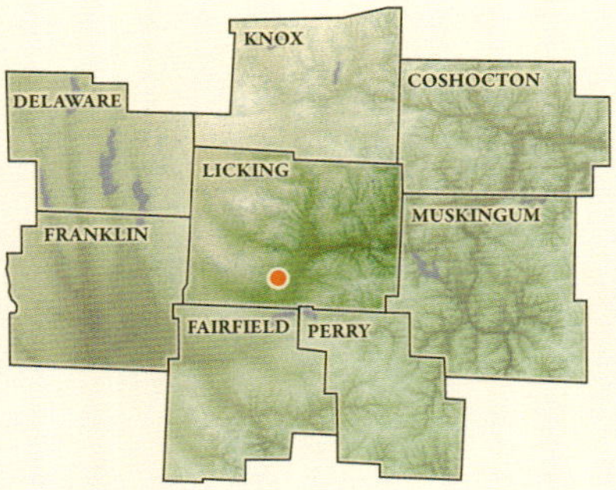

ACREAGE: 12

OWNERSHIP: Ohio Division of Natural Areas and Preserves

ACCESS: By permit only

WEB SITE RESOURCES: http://www.dnr.state.oh.us/dnap/location/cranberry.html

A need for improved commerce to service Ohio's burgeoning postsettlement population accidentally created Cranberry Island. The 1830s ushered in the canal era; waterways accommodating horse-drawn barges were hailed as the answer to moving goods around the state.

Around 1834, construction of Buckeye Lake began. Its purpose was as a feeder lake for the Ohio-Erie Canal, allowing the channel to cross the divide between Ohio River and Lake Erie watersheds. The valley where the lake now sets was the site of an extensive bog, and as the newly created waters of Buckeye Lake slowly rose after damming, the youngest, most buoyant part of the bog began to rise.

As the bog ascended like a botanical phoenix, it remained rooted to the substrate; thus, Cranberry Island doesn't really float. When it first surfaced, the island was about sixty-five acres; today it is perhaps twelve. A constant chemical clash between alkaline lake waters and acidic bog substrates melts away the island. A large tree occasionally topples along the edges, taking big chunks of the bog along. The island's lifespan isn't long, so see it while you can.

Cranberry Island is built on *Sphagnum* moss—peat—and these bryophytes exude acidity as a byproduct of their growth. Thus, the substrate is low in pH, holds water like a sponge, and creates refrigeratorlike root zone temperatures. Life in the bog is tough, and only specialized plants can thrive. One common plant every visitor should learn is poison sumac, which causes a blistering dermatitis.

Following a short boat ride from the mainland, visitors can easily traverse the bog via a boardwalk. The place lives up to its name—large cranberry blankets the meadows, so much so that locals once made cranberry-harvesting trips to the island. Pitcher-plants are everywhere and seem right at home here. However, they are thought to be the spawn of intentional introductions in the early 1900s.

To attract insect prey, pitcher-plants secrete aromatic compounds that lure victims into their pitcherlike leaves. Pitcher-plant juice is filled with proteins and nitrogen that allow the plant to grow in the harsh substrate of a peat bog. An American toad hides among these plants.

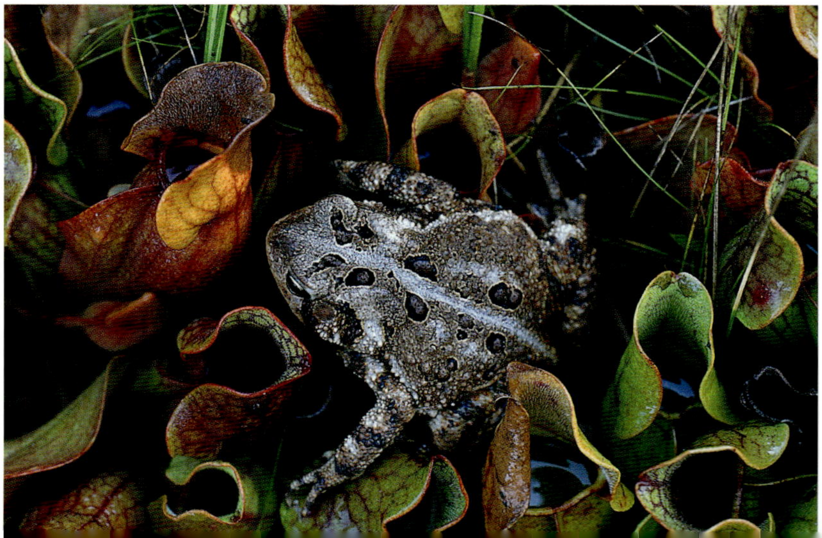

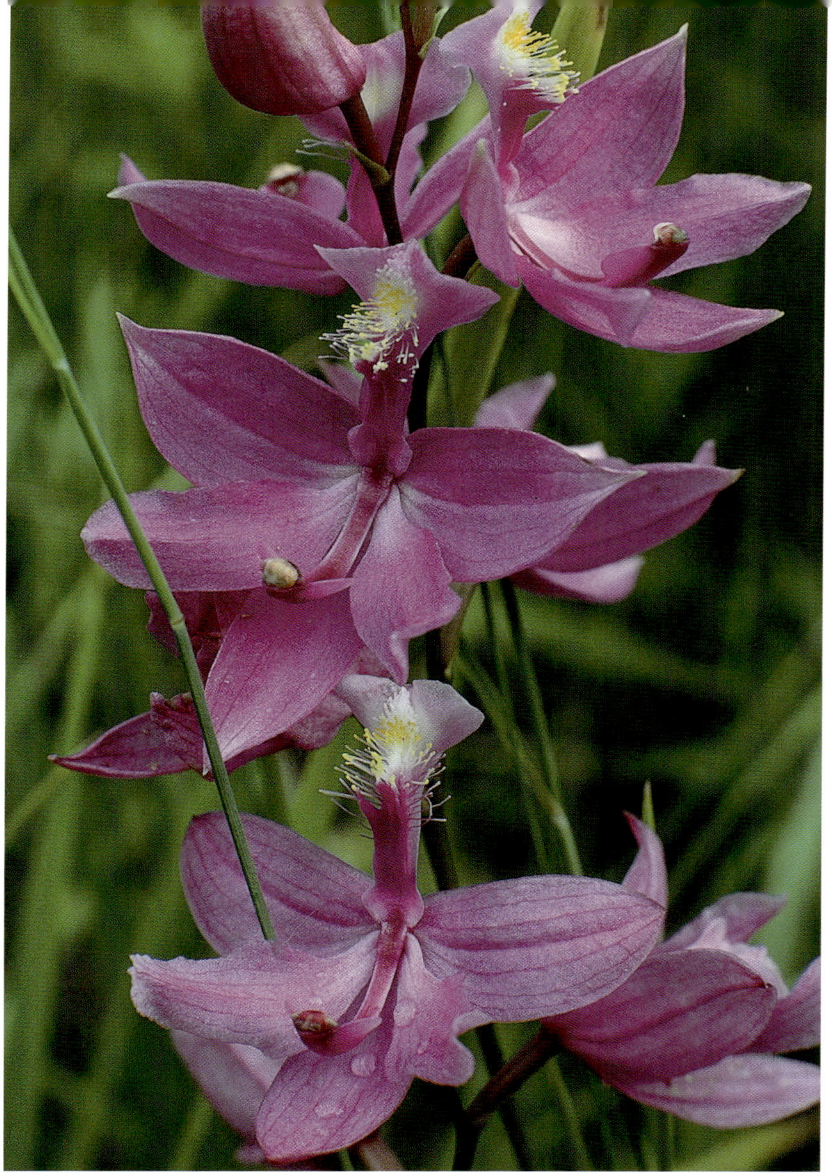

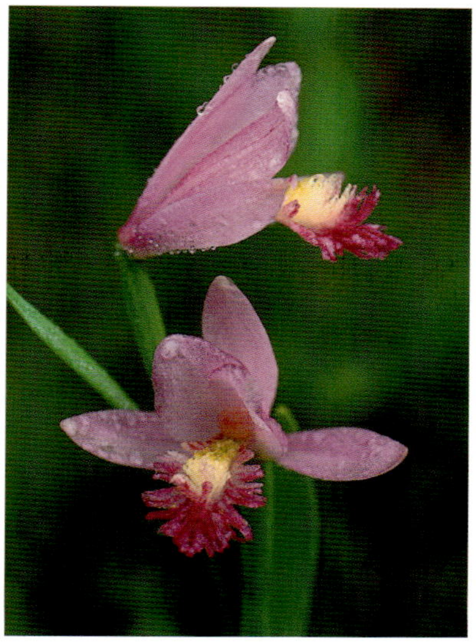

Left: Many orchids, grass-pinks included, lack nectar. To compensate, grass-pinks rely on botanical subterfuge to attract insect pollinators.

Above: The rose pogonia is one of forty-six species of native Ohio orchids. Addicted to acid soils, this striking species begins to flower in June.

Mid- to late June may be the best time to visit. That's when bright pink blooms of two rare orchids, rose pogonia and grass-pink, enliven the meadows. The latter lures bees with an appealing brush of hairs on the upper lip. Seeking nectar, the insect alights and is promptly catapulted downward onto the column and is dusted with pollen. The bee, duped, gets nothing in return.

When naturally formed, bogs usually surround lakes, eventually filling them in with vegetation. Here, the opposite has occurred—possibly the only such site anywhere. While "unique" is an often misapplied and overused descriptor, truly it is accurate in regard to Cranberry Island.

Killdeer Plains/ Big Island Wildlife Areas

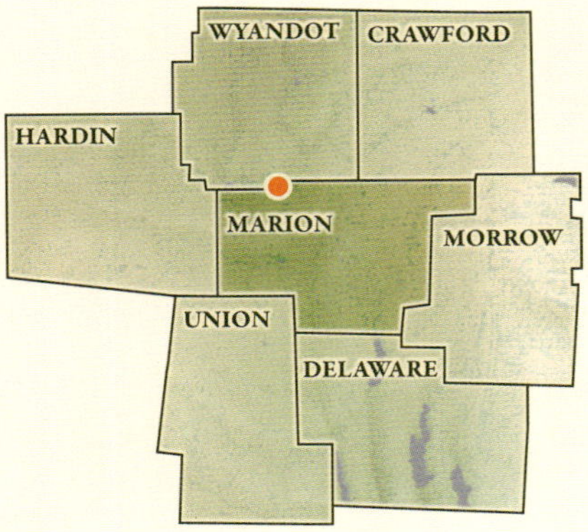

ACREAGE: 13,659

OWNERSHIP: Ohio Division of Wildlife

ACCESS: Open to the public except for signed restricted zones

WEB SITE RESOURCES:

http://www.dnr.state.oh.us/Home/tabid/4414/Default.aspx

http://www.ohiobirds.org/birding-sites/countymap.php

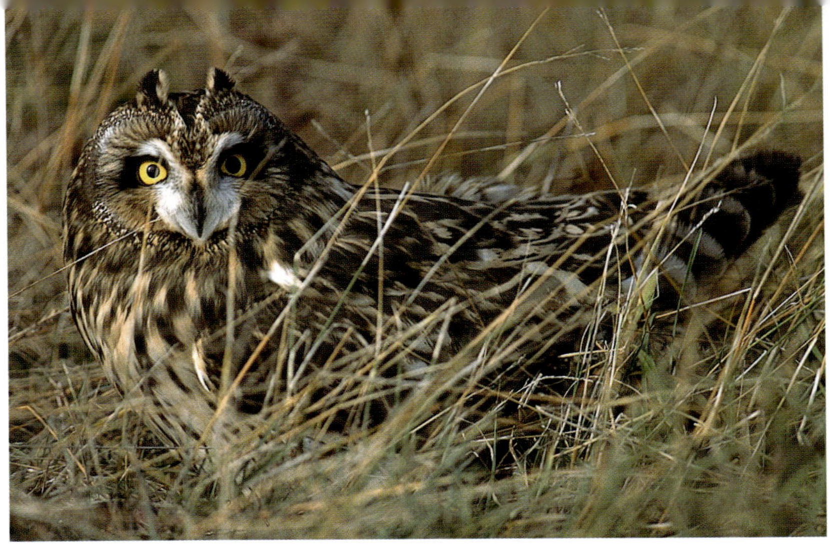

Above: Short-eared owls were no doubt a common breeder in Ohio's original prairies. Killdeer and Big Island can be productive locales for these grassland owls.

Opposite bottom: The loss of more than 90 percent of Ohio's original wetlands has greatly reduced populations of water birds. The least bittern is now listed as threatened in Ohio. Small numbers of this heron species breed at Big Island and Killdeer.

One of Ohio's largest prairie expanses was the Sandusky Plains, which covered areas of Crawford, Hardin, Marion, and Wyandot counties. Until settlement, this mosaic of wetlands and oak savannas covered 192,000 acres. By far the largest protected piece of the Sandusky Plains is within these two wildlife areas, although they conserve but 7 percent of the original landscape.

Botanical remnants of our prairie past are especially evident at Killdeer Plains. Wet swales and roadside ditches support stands of prairie cord grass, wheat sedge, Riddell's goldenrod, and prairie-dock. One of about one hundred plants originally discovered in Ohio, the goldenrod was first found in the 1830s by pioneer botanist John Riddell and named in his honor.

A great rarity is rope dodder. An odd morning-glory relative, it lacks chlorophyll and forms thick braidlike strands resembling orange pipe cleaners. Deriving nutrients by parasitizing host plants, it usually attacks Canada goldenrod or sawtooth sunflower. Heat from prairie fires is apparently necessary to crack the dodder's thick seed coats and allow germination.

Grassy meadows at Killdeer provide refugia for three of Ohio's rarest snakes. Small and ribbonlike, the plains garter snake is only known from Killdeer Plains. This relict colony is the easternmost known population. Eastern Massasauga rattlesnakes also inhabit wet fields and remain fairly common—Killdeer is one of their best Ohio strongholds. Stunning lime-colored smooth green snakes,which seldom reach two feet in length and are quite docile, also share the grasslands.

Big Island is mostly wetlands, which host myriad waterfowl, nesting bald eagles, migrant shorebirds, and rare breeding marsh birds. Three wetland cells at the south side of the area total nearly four hundred acres. Ducks and geese pack these marshes in spring and fall, and rare Ohio nesters such as American and least bitterns, king rail, redhead, and Northern pintail sometimes breed. Lush oases in seas of cropland, Big Island and Killdeer have lured many rarities, to the excitement of birders. Notables include long-billed curlew, Northern wheatear, white ibis, and black-necked stilt. The winter season brings roosts of long-eared and short-eared owls to Killdeer Plains. Raptors such as Northern harrier and rough-legged hawk become common.

Although most prairie sod in the Sandusky Plains has been tilled and planted to beans, corn, and wheat, birds that evolved migrations that correlate with prairie ecosystems still sweep through on great passages. Muddy fields of black prairie humus grow verdant crops but still lie bare in early spring. In April, large flocks of American golden-plover pass through; sometimes over a thousand whirl about fields delivering melodious haunting whistles. Returning from Argentinean pampas, they are headed to Arctic breeding sites, and the Sandusky Plains provide critical resting stops.

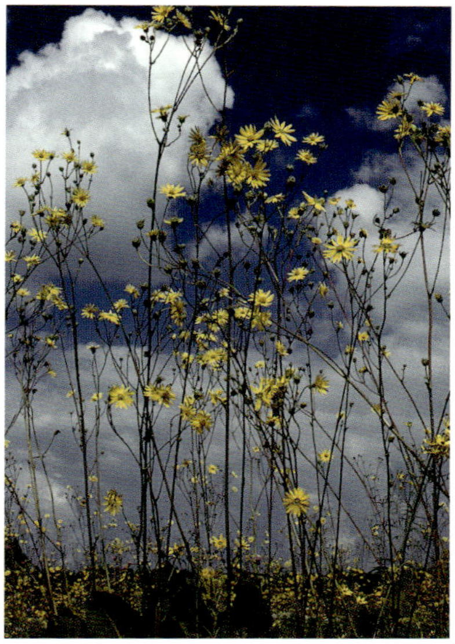

Killdeer and Big Island are part of the formerly vast Sandusky Plains, which historically blanketed thousands of acres. Virtually the entire original prairie is now farmland. Prairie plants such as prairie-dock still persist at Killdeer.

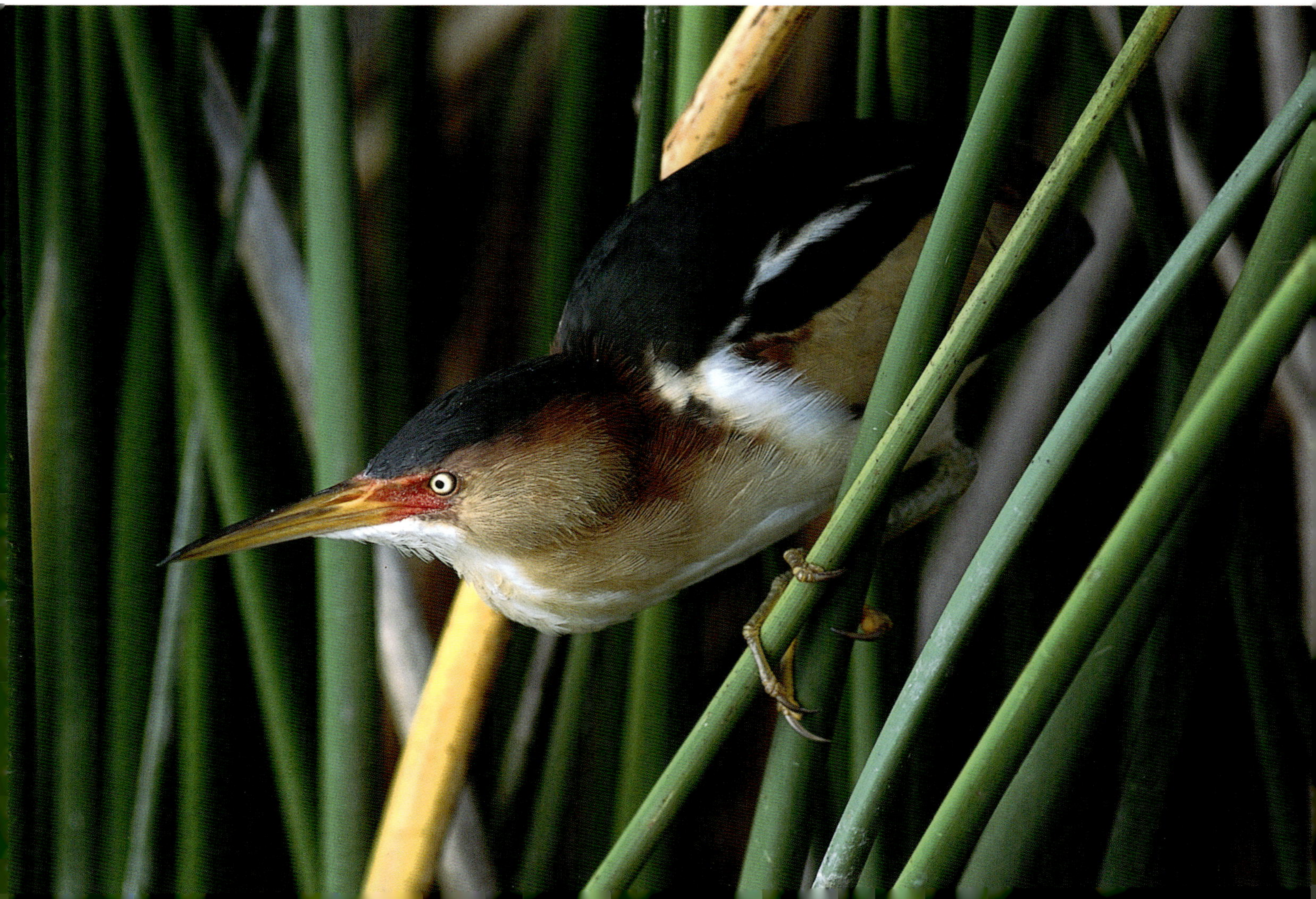

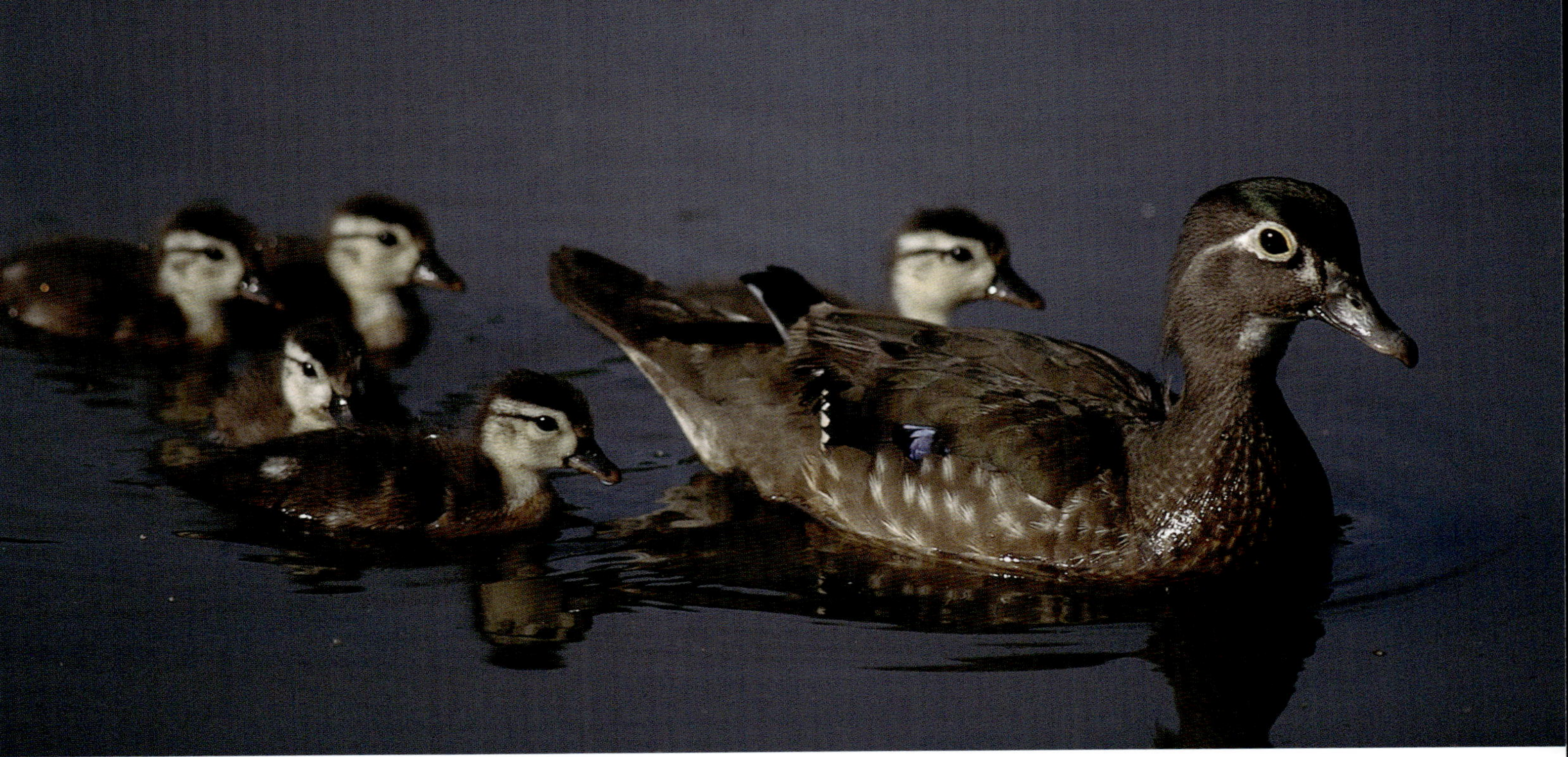

Few animals better represent successful conservation in North America than the wood duck. Because of rampant overhunting and habitat loss in the nineteenth and early twentieth centuries, populations plummeted to the point where some ornithologists were predicting possible extinction by 1930. Increased protection of wetlands, regulation of hunting, and erection of nest boxes turned the tables, and today wood ducks are once again common.

Recent wetland restorations at Big Island have demonstrated the reclamation potential of former prairie. In the late 1990s the hydrology of hundreds of acres of former wetlands was revived by busting drainage tiles and constructing berms to trap water. Within two years, blue-winged teal, American coot, pied-billed grebes, and other water birds were nesting. In 2000, a pair of Wilson's phalarope, a very rare prairie breeder, nested, then returned and bred again in 2002. This project showed that with a bit of creativity, prairies and the animals that use them can bounce back.

Rescued in the nick of time, many large trees in Lawrence Woods were already marked with paint to indicate their imminent harvest when the site was acquired in 1997. Because the Ohio Division of Natural Areas and Preserves successfully intervened, Ohioans can enjoy the largest protected old-growth woodland in western Ohio.

Prior to European settlement, woodlands like this were commonplace. Excepting some prairie regions, much of the western Ohio till plains were blanketed with forest, ranging from beech-maple and oak-hickory associations on higher grounds to elm-ash-dominated swamps on poorly drained sites.

A trip around the Lawrence Woods boardwalk offers a historical perspective into the vast forests that once covered this region. Elevated knolls, although only a few feet higher than the lowlands, support upland trees such as shagbark hickory and white oak. An interesting example of the latter occurs near the trail entrance. Twisted from some long-ago disturbance, this ancient white oak's distorted trunk creates an almost perfect silhouette of a rhinoceros.

The lowest spots hold water and are choked with buttonbush, an aquatic shrub. Buttonbush swamps are pools of diversity within forest ecosystems, and the large one at Lawrence Woods is an outstanding example. Copious white blooms of buttonbush attract legions of pollinators: bees, moths, butterflies, and others. Tranquil black waters of buttonbush pools are usually fishless; thus, aquatic insect and amphibian larvae survival soars. Perhaps the most interesting amphibian

Lawrence Woods State Nature Preserve

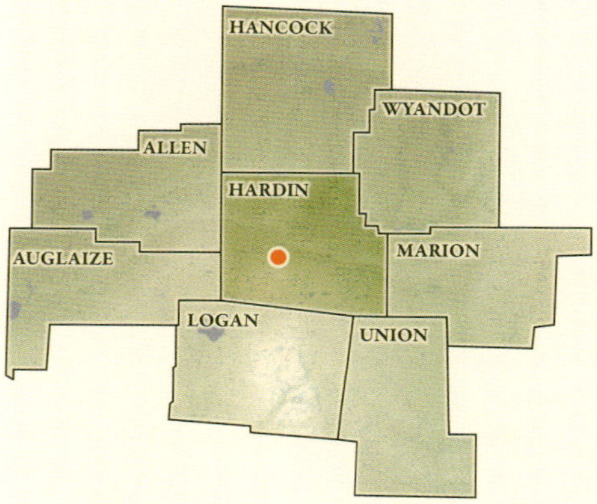

ACREAGE: 1,035

OWNERSHIP: Ohio Division of Natural Areas and Preserves

ACCESS: Open to the public during posted hours

WEB SITE RESOURCES: http://www.dnr.state.oh.us/dnap/location/lawrence_woods/tabid/905/Default.aspx

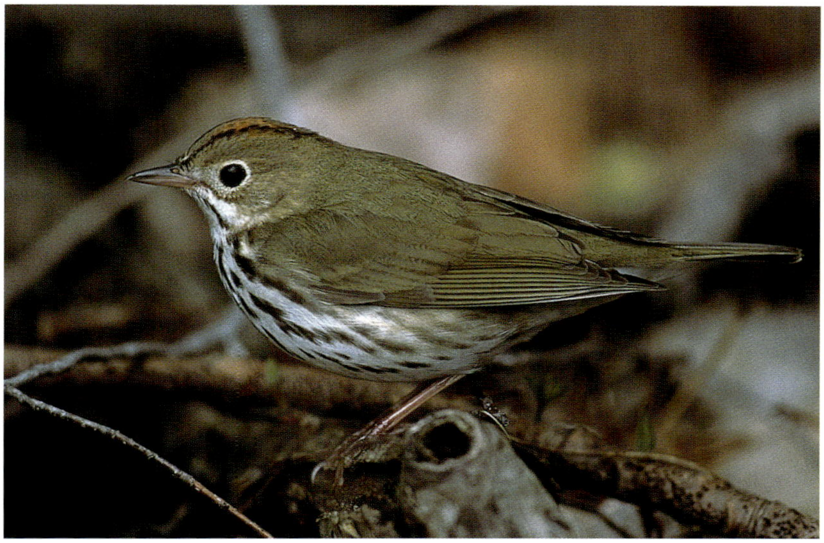

A loud, ringing *Teacher, Teacher, Teacher* announces the ovenbird. This ground-feeding warbler walks about the forest floor gleaning through leaf litter. The nest is placed on the ground and resembles an old-fashioned Dutch oven.

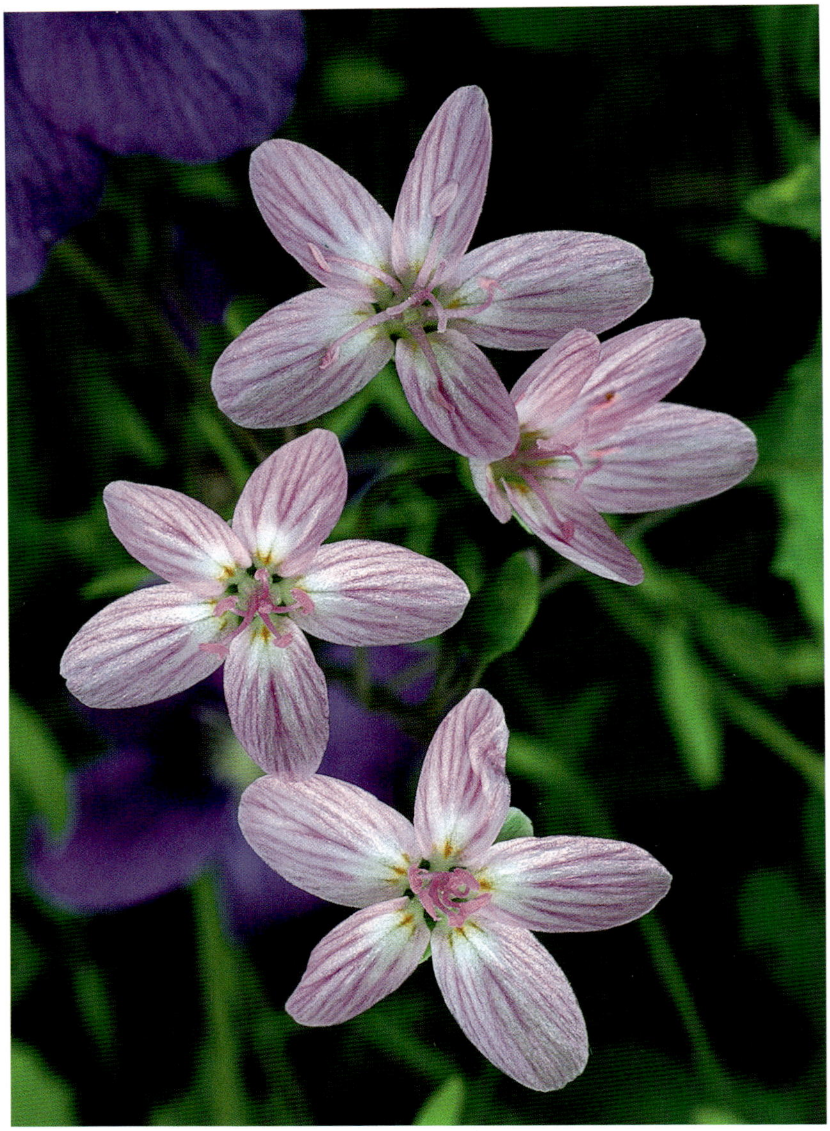

Spring-beauties carpet Ohio woodlands in April with boldly striped pink flowers. It and scores of other wildflowers provide a visual vernal feast in Lawrence Woods.

breeding here is the tiger salamander. A behemoth, an adult may reach a foot in length.

A summertime visit to Lawrence Woods' buttonbush swamp bears out the importance of this habitat for dragonflies. Myriad Eastern pondhawks, common green darners, black saddlebags, and blue dashers hunt, mate, and fight over the sheltered waters.

The soupy outflow of the buttonbush swamp forms a swampy wetland with a plant community unique in Ohio. A mucky quagmire sports over one thousand plants of the endangered heart-leaved plantain. Their large, cabbagelike leaves contrast with the strap-shaped fleshy leaves of tubercled rein orchid, which grows intermixed. Nowhere else in Ohio do these two rarities occur together.

Plenty of birds breed here, and woodpeckers are especially obvious. All six of Ohio's common nesting species are found. Large oval-shaped excavations from crow-sized pileated woodpeckers are conspicuous, and visitors are likely to hear the maniacal laughing calls of this giant, if not actually see the birds. Old pileated nest holes are particularly valuable to the forest ecosystem. Multitudes of other animals utilize these cavities, including Eastern screech-owl, gray and Southern flying squirrels, Virginia opossums, black rat snakes, and raccoons.

April and May are spectacular. The woodland teems with dozens of species of wildflowers in nearly every hue of the rainbow. Violets are especially eye-catching, and at least a third of the twenty-seven native species can be found. The forest floor is thick with other species, including various trillium and waterleaf, cut-leaved toothwort, wild geranium, wild blue phlox, and bloodroot.

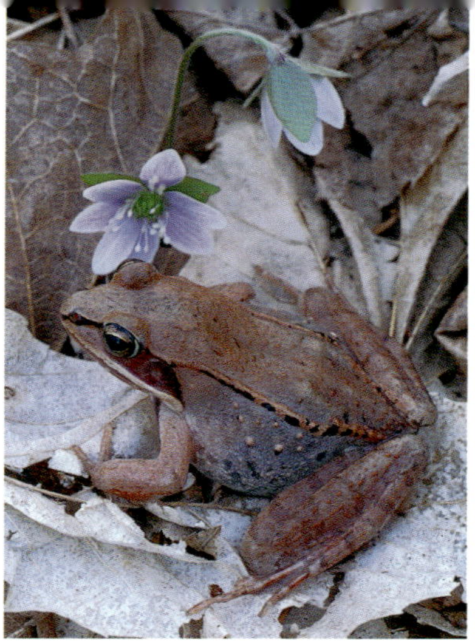

Above: Wood frogs congregate in shallow breeding pools. Their ducklike quacking call is distinctive. This one is with early blooming hepatica.

Below: With the first warm rains of spring, spotted salamanders migrate to vernal breeding pools, where females lay jellylike egg masses.

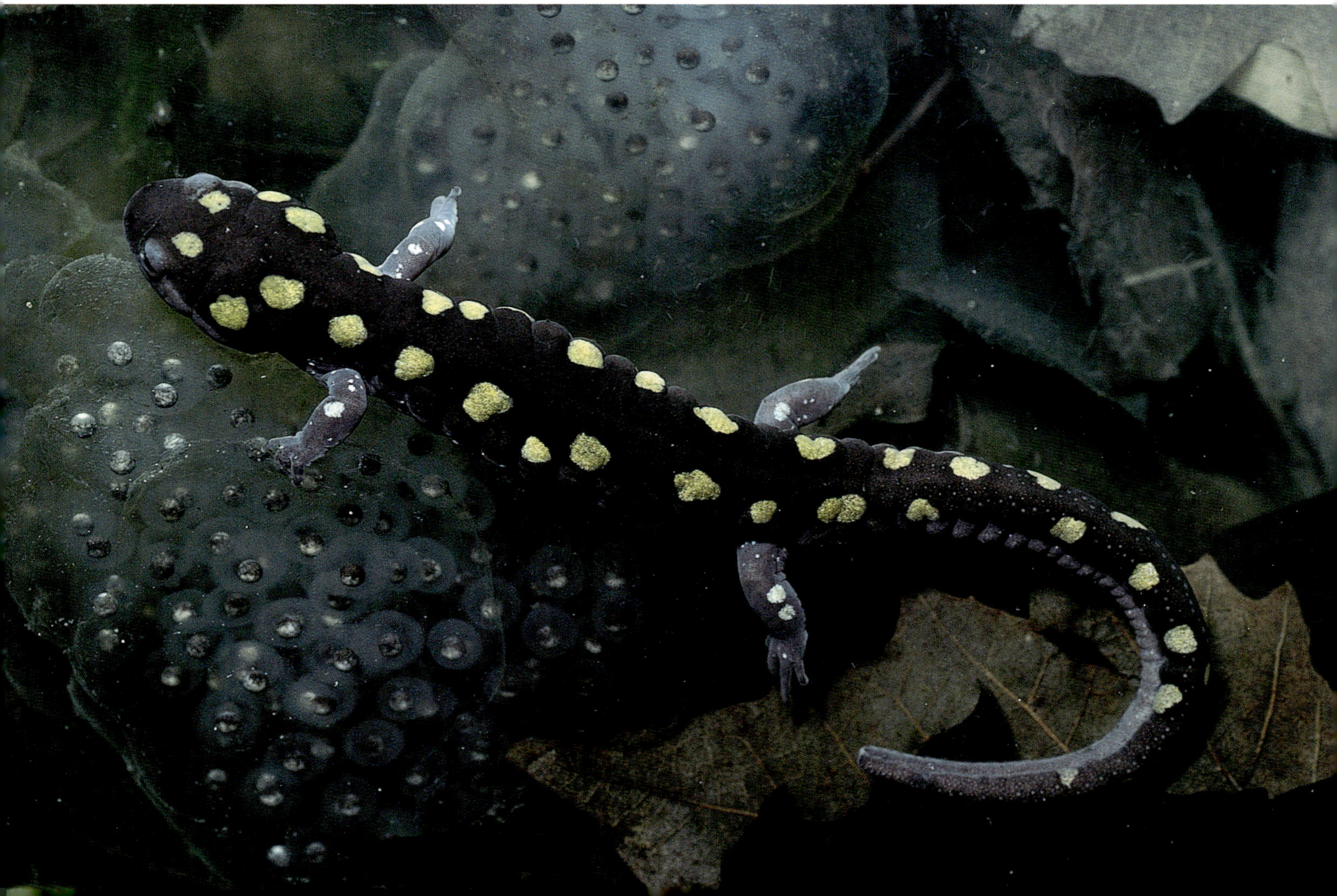

Mud Lake Bog State Nature Preserve

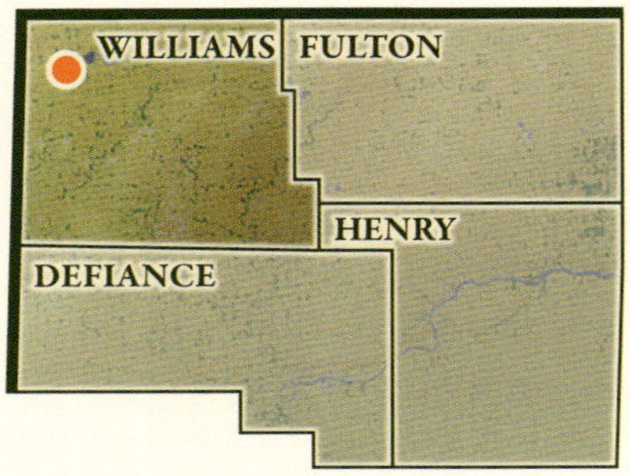

ACREAGE: 74

OWNERSHIP: Ohio Division of Natural Areas and Preserves

ACCESS: Access by permit only

WEB SITE RESOURCES: http://www.ohiodnr.com/dnap/permitonly/tabid/863/Default.aspx

Mud Lake is the best remaining example of a glacial lake in northwestern Ohio. There aren't many competitors. Williams County was once dotted with numerous kettle hole bogs, but most have succumbed to the inexorable forces of ecological succession. Only from the air in leafless winter can one get a true sense of the former abundance of bogs, as circular depressions grown over with vegetation dapple the landscape.

Almost perfectly round, Mud Lake is ringed with rare flora. The western side is dominated by tamarack underlain with luxuriant beds of *Sphagnum* moss. The eastern side is more fenlike; alkaline seeps foster vegetation characteristic of high pH soils. Everywhere things are wet, and visitors on foot must tread carefully or risk being sucked into the mire.

Although the lake and its associated habitats cover only sixteen acres, rare animals and plants abound, offering one of Ohio's densest concentrations of extraordinary species. Thirteen species of threatened or endangered plants grow here, and nine species of equally rare animals live in or around the lake.

The muck-bottomed lake's cold water harbors noteworthy aquatic plant communities. Rare floating and flat-stemmed pondweeds grow with threatened American water-milfoil, forming dense underwater tangles. Iowa darters and lake chubsuckers, fish species dependent upon excellent water quality and intact aquatic plant associations, live in the submerged foliage. Showy white blooms of fragrant water-lily hang on the water's surface, their circular floating leaves providing landing pads for one of Ohio's rarest damselflies, the lilypad forktail. This tiny insect is only known from Mud Lake and wasn't discovered here until 1992. On the other end of the dragonfly size scale are Canada and mottled darners, both endangered. These large mosaic darners are brightly colored with tints of blue, green, and yellow and occur no farther south.

Nearby swamp forests harbor the federally threatened copperbelly water snake. A bruiser, this fiery-tempered reptile can reach six feet in length. Like most snakes they are furtive, but this species is particularly difficult to locate as it occurs in wet, hard to access habitats and is quite secretive. Shiny coal-black above, and bright orange-red below, copperbellies are extraordinarily showy reptiles. Williams County habitats support some of the few remaining relict populations of this snake.

A botanist can't take two steps along the shore without finding a rare plant. Shrub thickets are populated with swamp birch and slen-

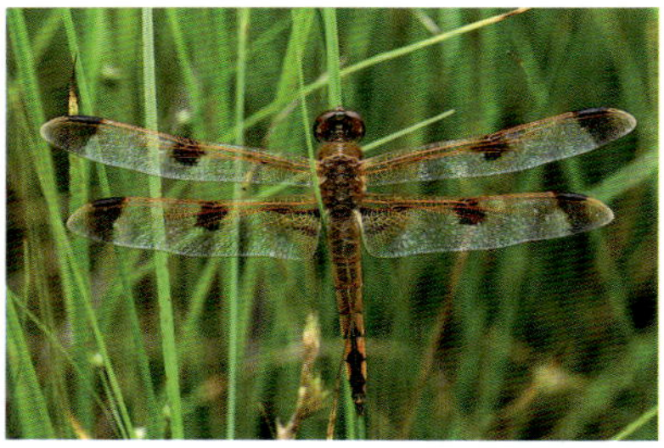

Above: Mud Lake Bog contains an outstanding odonate fauna. Beautiful dragonflies like this painted skimmer can be seen as well as rarer species like Canada darner.

Opposite top: Most of the 108 glacial lakes in Ohio have been degraded and no longer support high fish diversity. Lake chubsuckers require abundant aquatic plants, clear cold water, and clean lake bottoms.

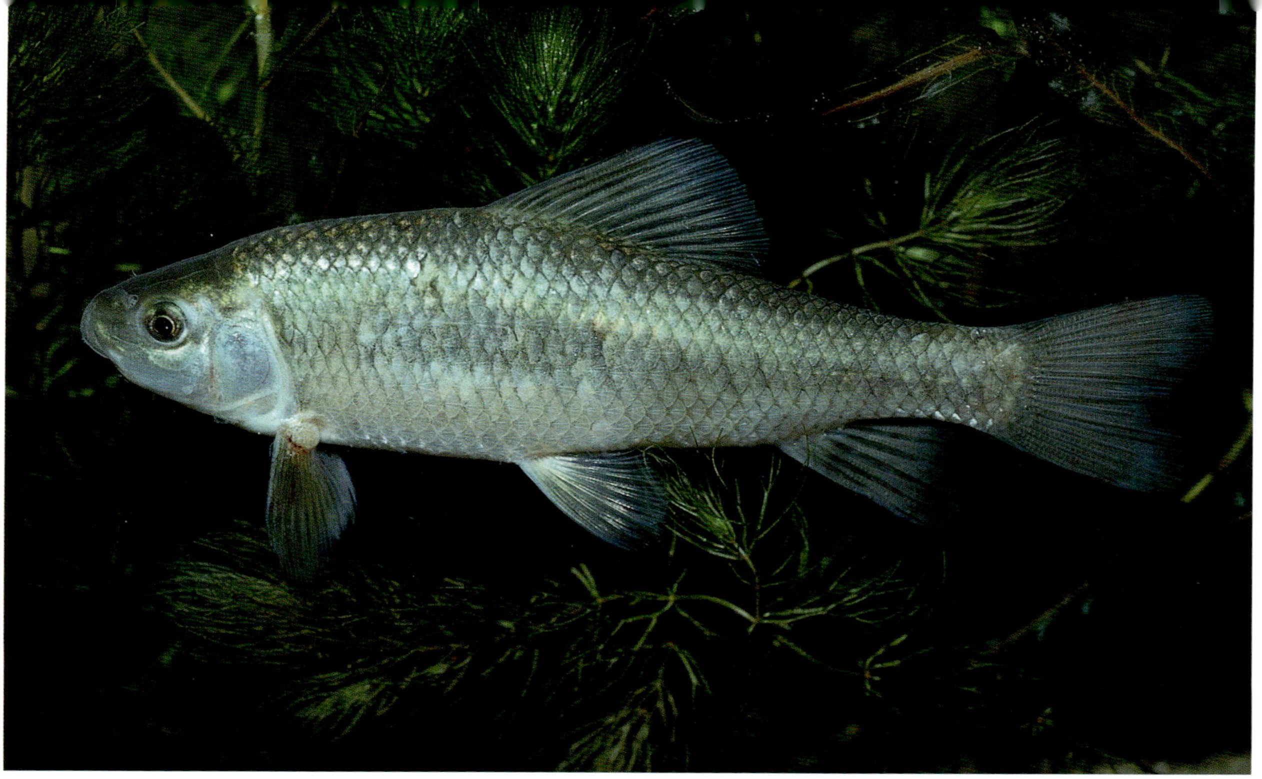

der willow, threatened and endangered, respectively. Boggy mats are rich with unusual flora, including one of the few Ohio stands of small purple foxglove. This diminutive figwort has stunning magenta blossoms prominently streaked with violet nectar guides to lure insect pollinators. Rarer yet is swaying-rush. This sedge of northern glacial lakes grows in deep lake waters, only its flowering tips protruding. Rooted in quicksandlike mire, it is physically impossible to reach on foot. Swaying-rush is one of nine state-listed rare sedges growing at Mud Lake.

Protection of the waters that flow into glacial lakes is a key to their conservation. Many aquatic organisms that inhabit pristine habitats such as Mud Lake are sensitive to water-quality degradation. Fortunately, the Ohio Division of Natural Areas and Preserves has protected much of Mud Lake's drainage, ensuring that future generations can study the flora and fauna of this fascinating ecosystem.

Below: Breeding primarily in peatlands, alder flycatchers are local in Ohio. Alder and the more common willow flycatcher can only be identified by song.

Overleaf: An endangered species, purple fringed orchids are known from perhaps six Ohio sites, with few individuals present in each population. Collectively, there may be fewer than one hundred plants in the state.

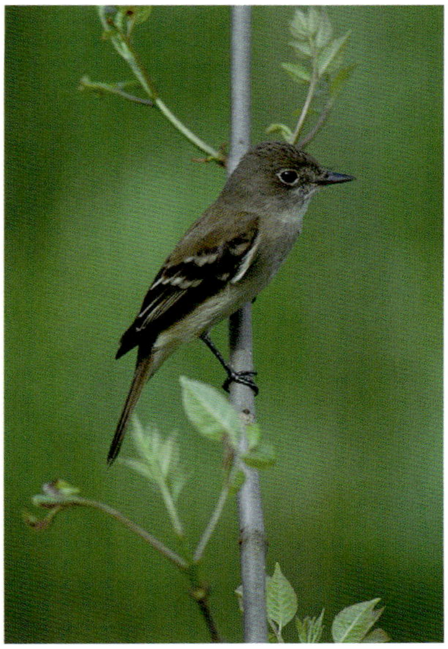

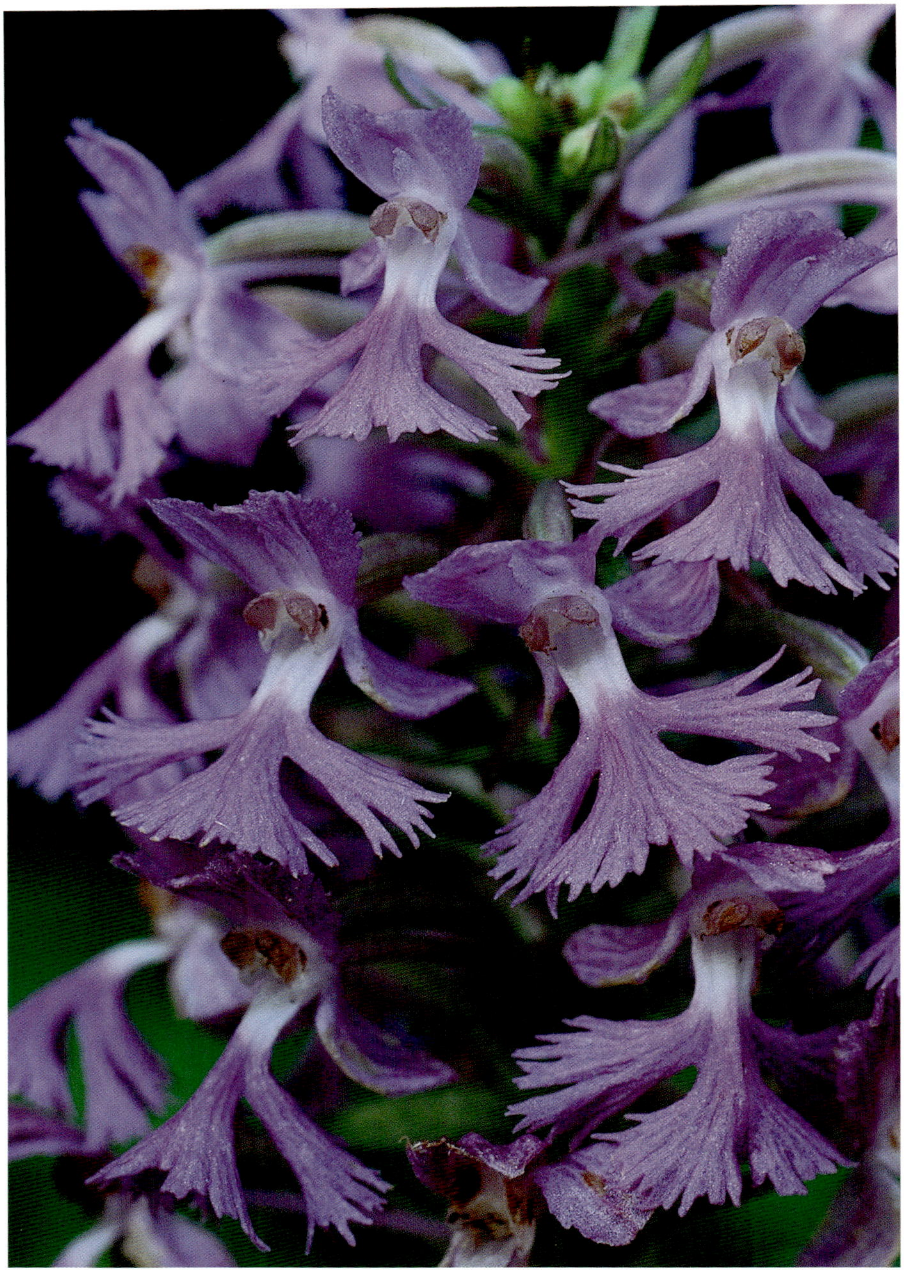

Huron-Erie Lake Plains

Forrest Woods Nature Preserve (1) • Goll Woods State Nature Preserve (2) • Headlands Dunes State Nature Preserve (3) • Irwin Prairie State Nature Preserve (6) • Kelleys Island (5) • Kitty Todd Preserve (4) • Magee Marsh Wildlife Area/ Ottawa National Wildlife Refuge (7) • Oak Openings Preserve (8) • Resthaven Wildlife Area (9)

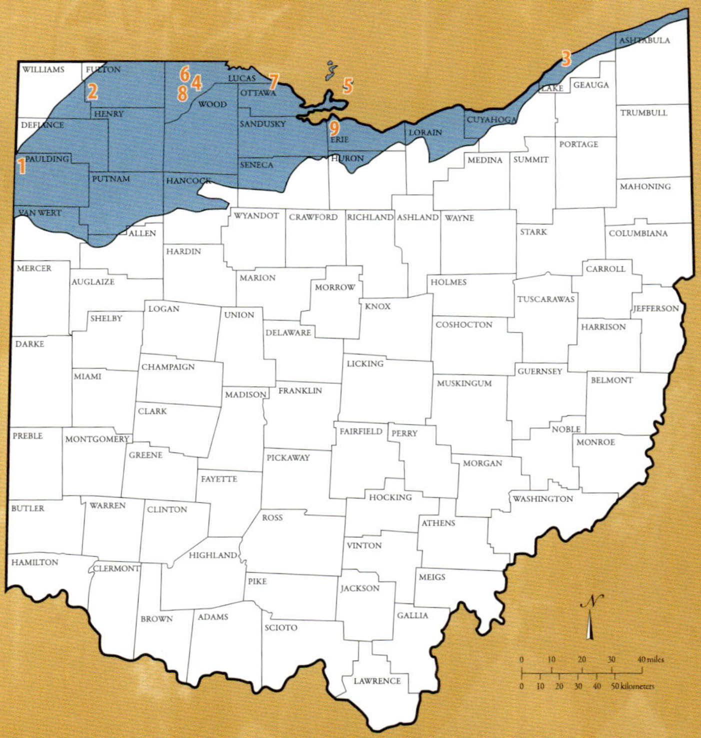

Forrest Woods Nature Preserve

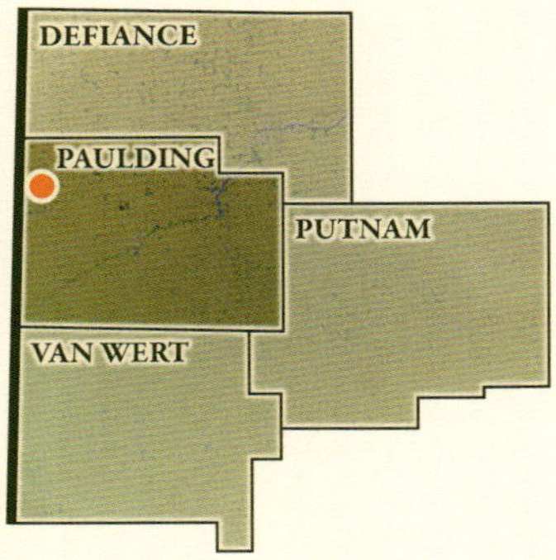

ACREAGE: 257

OWNERSHIP: Black Swamp Conservancy

ACCESS: By permit only through Black Swamp Conservancy

WEB SITE RESOURCES: http://www.blackswamp.org/

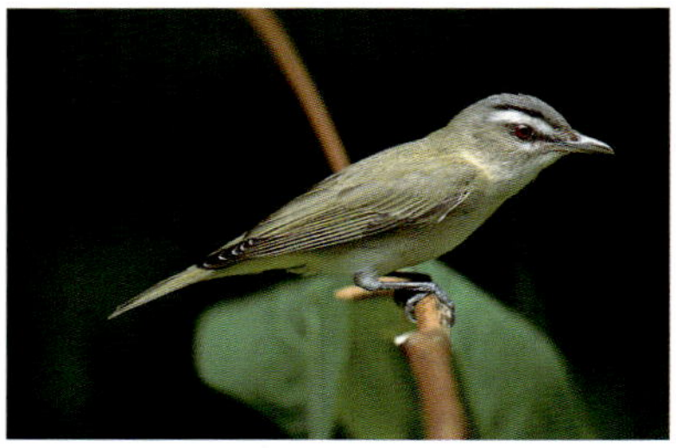

The vociferous red-eyed vireo is the most numerous neotropical migrant breeding in the deciduous forests of eastern North America. One male was documented singing over twenty-two thousand songs in one day. Territories can be as little as an acre or two, so even small woodlands can host several pairs.

Rivers are America's arteries, distributing plants and animals throughout the landscape. Marie DeLarme Creek, a tributary of the Maumee River, flows through the center of Forrest Woods. Near Fort Wayne, Indiana, a low-lying glacial outwash terrace has provided a natural conduit from Mississippi River drainages into the Great Lakes. For eons, southern species have migrated through this gap and infiltrated western Ohio as a result. Some of them occur in Forrest Woods, perhaps the best remaining riparian woodland on the Till Plains.

Forrest Woods lies on the southern periphery of what was once Ohio's greatest expanse of wetlands—the Great Black Swamp. The area stretched from present-day Lake Erie west to Indiana and south to the Maumee River, and early settlers avoided it like the plague. It wasn't until the 1820s when the first roads were hacked through the swamp, but by the late 1800s nearly the entire quagmire had been drained and transformed to croplands.

Visitors to Forrest Woods, especially during spring floods, will experience the daunting landscape faced by Great Black Swamp trailblazers. Towering wetland trees like green ash, swamp white oak, and American basswood shade the swampy oxbows. Part of the package are mosquitoes, sometimes in dense clouds. Braving the insects is worth it—this is one of Ohio's only remaining old-growth tracts of river bottom woods in western Ohio.

Heavy spring rains inundate the bottomland swamp forests. Standing water pools for weeks in oxbows and depressions, creating habitat for specialized plants such as raven-foot sedge and leafy blue flag, both rare in Ohio. Botanists found an odd parasitic morning-glory relative here, cuspidate dodder, a first for Ohio and far removed from its western range.

Forrest Woods is a vital refuge for breeding birds, as most of the great swamp forests that covered western Ohio no longer exist. Birds like yellow-throated warbler, Louisiana waterthrush, and cerulean warbler nest here but are rare elsewhere in the region.

Protection of Forrest Woods has ramifications far beyond conserving this woodland and Marie DeLarme Creek. As the stream flows into the Maumee River, its water quality directly influences that of the Maumee, one of Ohio's officially designated scenic rivers. The Maumee is also Lake Erie's largest tributary, so its waters have a direct impact on the health of our greatest lake.

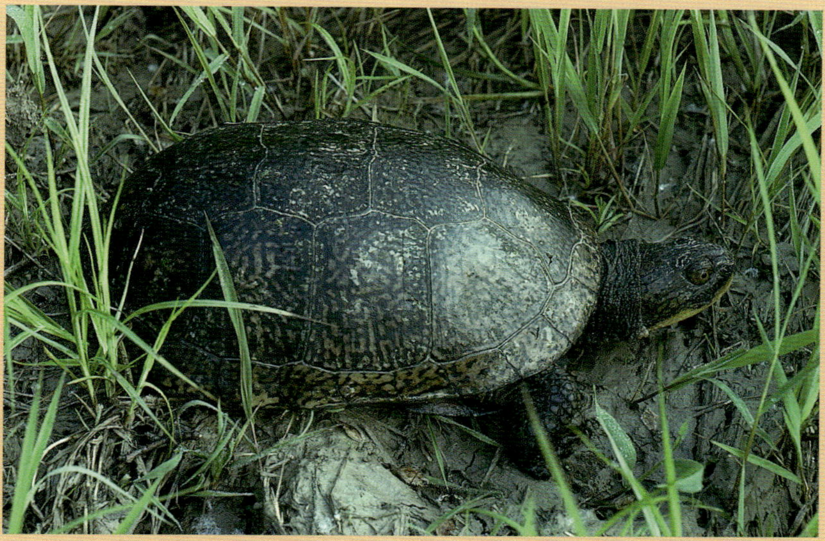

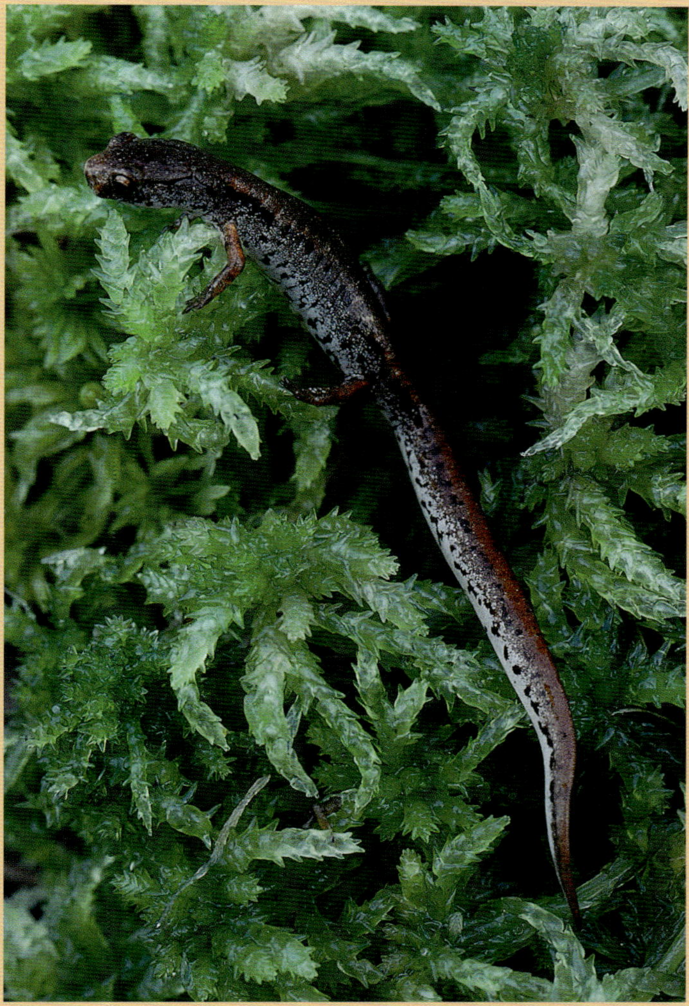

Top: Large and handsome, the Blanding's turtle can easily be recognized by its high-domed carapace and bright yellow throat. Never common, this species has a rather limited range. The Forrest Woods population is the southernmost known in the state.

Left: One of twenty-four species of salamanders in Ohio, the secretive four-toed salamander has been recorded from widely scattered locales throughout the state. Four-toeds breed in forested wetlands. Their eggs are laid under a cloak of moss, and adults remain to guard them until they hatch.

Goll Woods State Nature Preserve

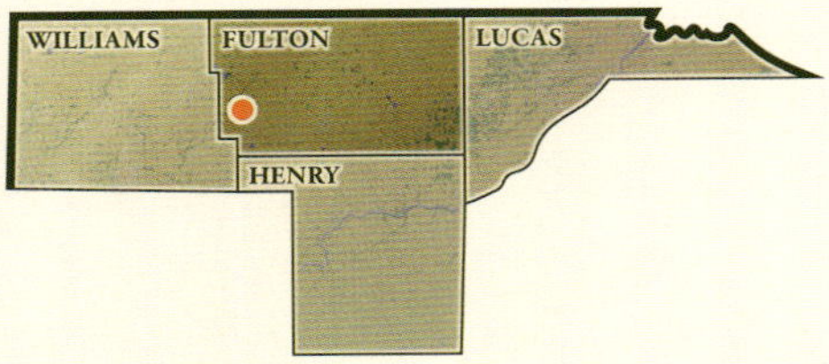

ACREAGE: 321

OWNERSHIP: Ohio Division of Natural Areas and Preserves

ACCESS: Open to the public during posted hours

WEB SITE RESOURCES:
http://www.dnr.state.oh.us/dnap/location/goll_woods.html

http://www.ohiobirds.org/birding-sites/showsite.php?Site_ID=56

Among the finest stands of old-growth timber remaining in Ohio, mammoth bur and red oaks, black ash, and tulip tree form a woodland cathedral at Goll Woods. The eighty-acre big woods are about all that's left of the Great Black Swamp, which blanketed much of the northwestern corner of the state well into the 1800s.

About 150 years ago, the region of Goll Woods was a wild place full of giant trees. The landscape was waterlogged and seemingly impenetrable. Animals long since vanished from Ohio were common: gray wolves, lynx, cougar, and black bear. An intrepid pioneer, Peter Goll, was one of the first to settle in this area, buying the land now occupied by Goll Woods in 1837 for $1.25 per acre.

Fortunately, the Goll family preserved a chunk of old-growth woods as nearly all the rest in the region fell to the axe and was converted to fields of beans and corn. In 1966 the Ohio Department of Natural Resources acquired Goll Woods, and today it is a state nature preserve.

Some of the trees stand over 120 feet tall and exceed twenty feet in circumference. Such a goliath can remove up to fifteen pounds of airborne pollutants annually and put thirty pounds of oxygen back into the atmosphere. Giant oaks also produce bumper crops of acorns. The nuts provide forage for gray squirrels, wild turkey, white-tailed deer, and others.

Spring wildflowers flourish in April and May, before leafy canopies shade the forest floor. Carpets of large-flowered trillium, cut-leaved toothwort, and bloodroot blanket the forest floor. Summer brings dense shade and fewer wildflowers, but one plant is especially noteworthy. One of few Ohio populations of the threatened three-birds orchid grows around the bases of massive American beech trees. This tiny orchid is largely subterranean and may only send stems and flowers above ground once every several years.

No Ohio forest fragment can truly be considered virgin. All have suffered some intrusion by the hand of man, and this is true of Goll Woods. As World War II accelerated and America became increasingly engaged, demand for timber to build ships escalated. Some of the huge trees in Goll Woods were harvested and their wood processed to provide flooring and other battleship accoutrements. Luckily, most of the sylvan giants have been spared for visitors to enjoy.

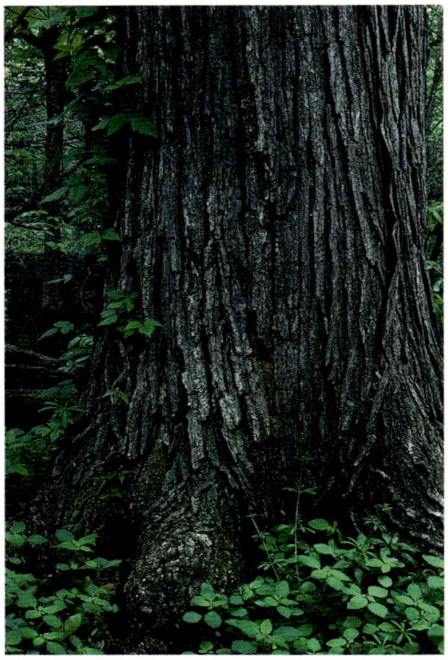

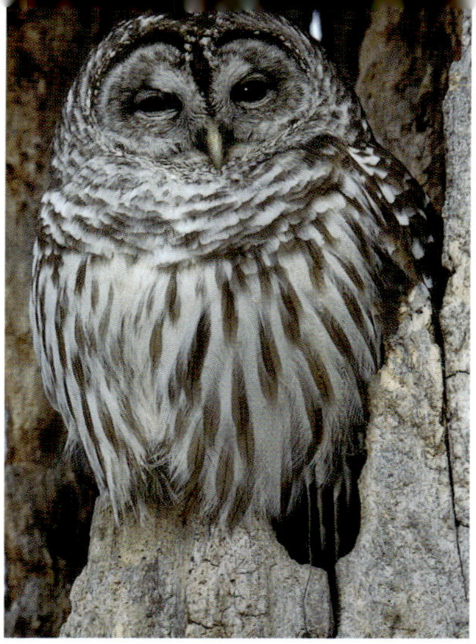

Opposite: An outstanding feature of Goll Woods is the abundance of bur oaks. Some magnificent old trees in this old-growth forest are four feet in diameter.

Above: One of three species of common, widespread breeding owls in Ohio, barred owls are strictly woodland dwellers. They do best in mature, swampy forests or ravines, and are most often detected by their calls.

Right: A delicate member of the buttercup family, hepaticas are often characteristic of rich woods, where they occur in association with many other native wildflowers.

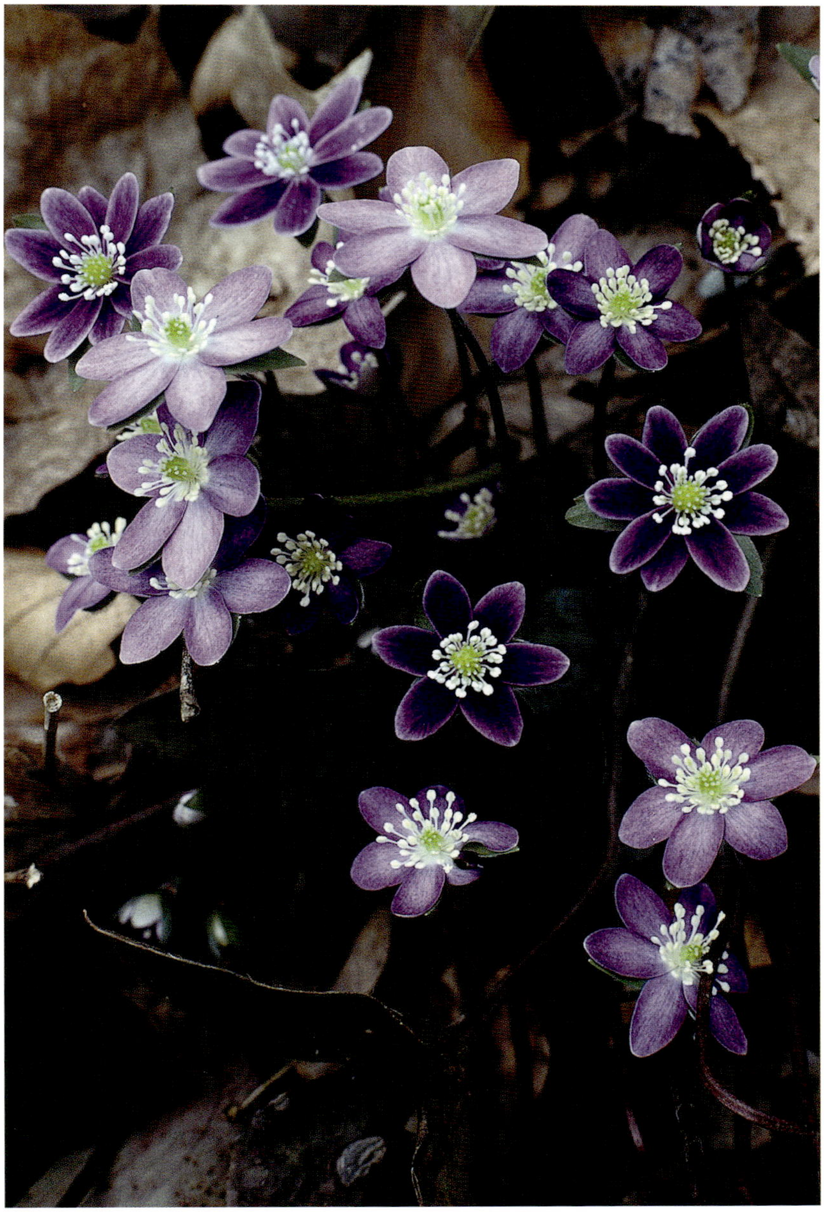

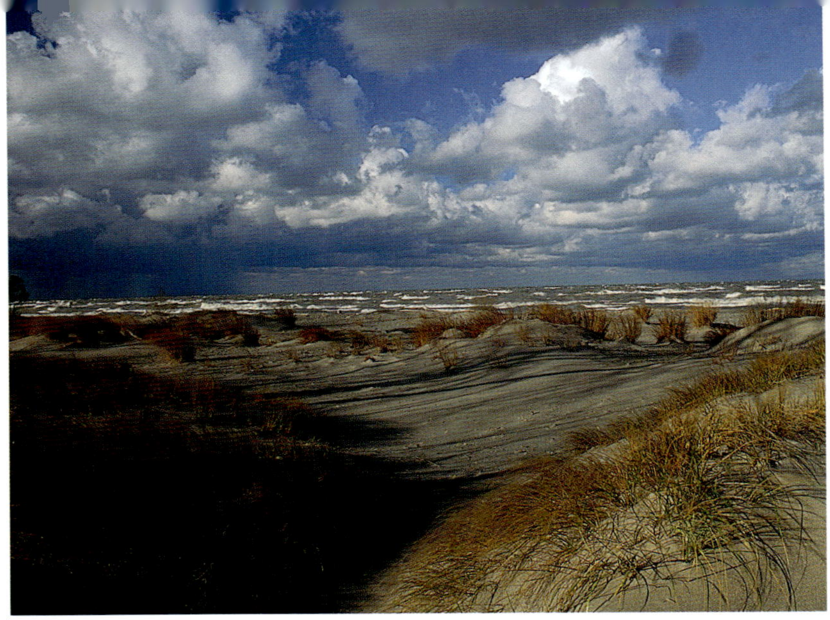

Headlands Dunes State Nature Preserve

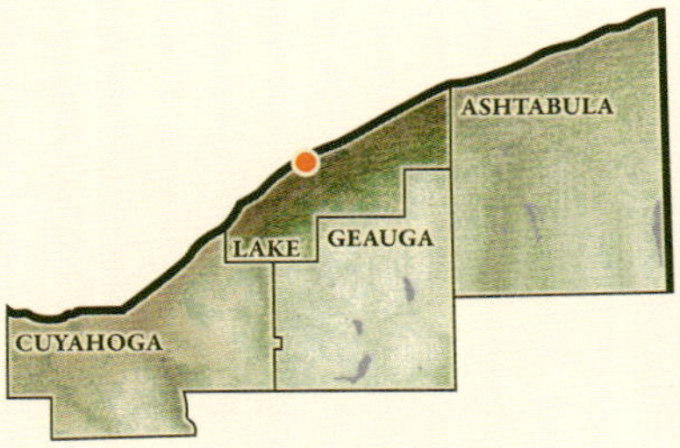

ACREAGE: 25

OWNERSHIP: Ohio Division of Natural Areas and Preserves

ACCESS: Open to the public during posted hours

WEB SITE RESOURCES:
http://www.dnr.state.oh.us/dnap/location/headlands.html

http://www.wcasohio.org/mentor_headlands_state_park.htm

Top: Headlands Dunes preserves one of the last and best beach dune communities in the state. Many coastal plain plant species survive here.

Opposite: Inland sea rocket is one of a small suite of Lake Erie beach plants that are Atlantic Coastal Plain disjuncts. These beachdwellers likely migrated westward via an interconnected band of lakes and wetlands that stretched east-west along the face of the retreating Wisconsin glacier some twelve thousand years ago and gradually became isolated.

Everybody loves a beach. That's been the downfall of most of Lake Erie's natural shoreline habitats—they're overrun with recreational users. At one time, over 170 miles of unsullied coastline stretched along Ohio's north coast; today, Headlands Dunes harbors the best of the few remaining lakefront plant communities.

The low sand dunes of Headlands, carpeted with thick stands of grasses, stand in stark contrast to the state park beach immediately to the west. There, most evidence of plants is eradicated—victims of decades of unrestricted trampling by visitors.

Although abundant at Headlands, two of the dominant grasses are rare in Ohio, confined strictly to a few isolated, undisturbed beaches. American beach grass forms lush stands on the outermost dunes. The tough rhizomes of this grass provide vegetative reinforcement to loose sand, preventing erosion. Just alee in sheltered swales are blankets of coastal little bluestem, another lakefront rarity.

Like botanical museum pieces, rare Atlantic coastal plain disjuncts grow on the wave-washed forebeach. Migrants of long ago, when Lake Erie was connected to the Atlantic seaboard by a band of postglacial wetlands, a handful of species persist in the Great Lakes. One of the most interesting is seaside spurge. This diminutive annual grows pancake-flat in sands just above the waterline and quickly exploits new habitats in ever-changing conditions.

Headlands is legendary among birders. Over 310 of the 418 bird species thus far recorded in Ohio have been found here, including many rarities. Headlands' location, coupled with good foraging and resting habitat, conspires to create a migrant trap perhaps unequaled elsewhere in the state. Large fallouts of songbirds occur routinely in

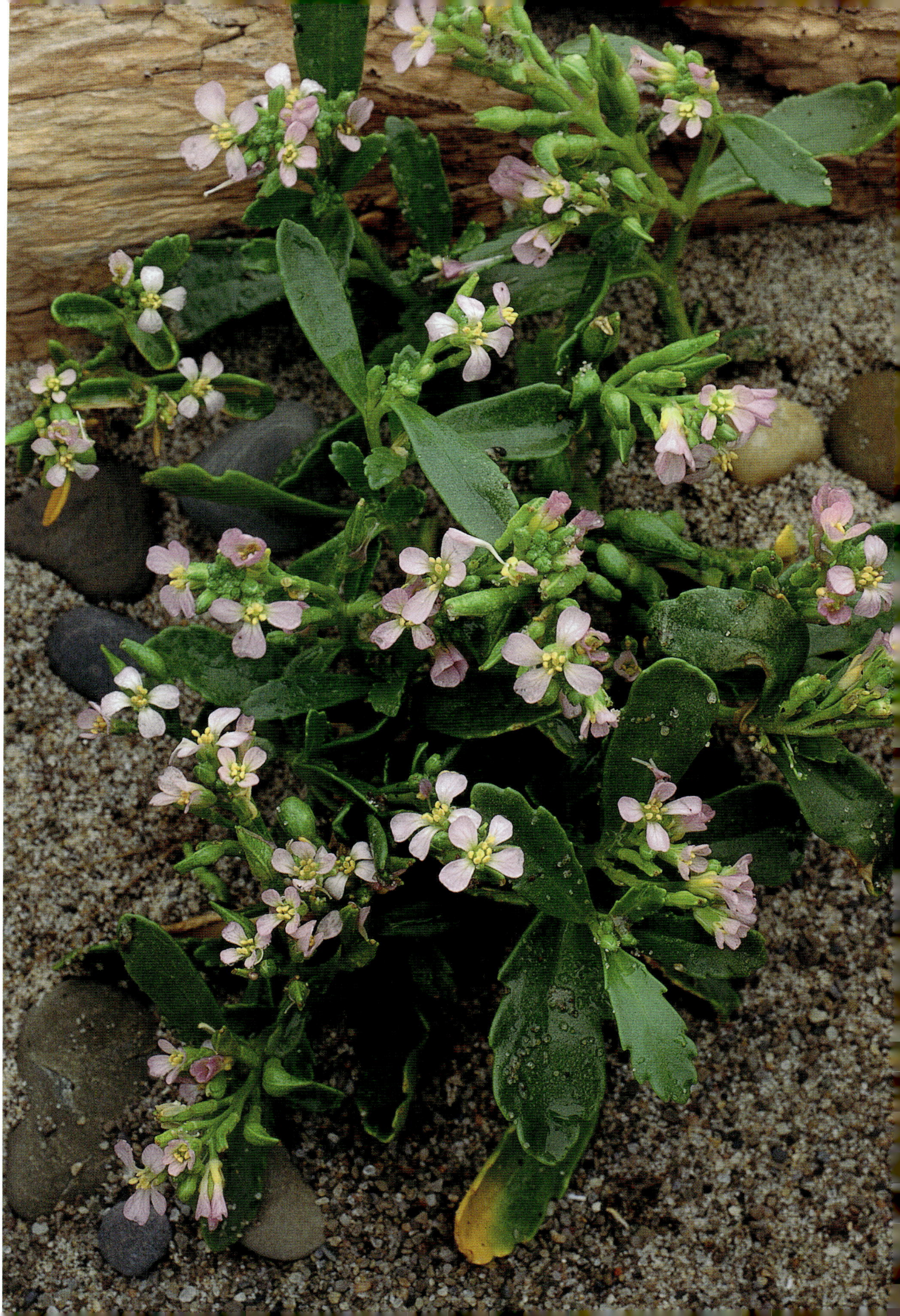

spring and fall; it's not unusual to tally over twenty-five species of warblers in a morning.

Waterbirds galore are attracted to Headlands' beach, the sheltered harbor of the Grand River (Fairport Harbor), and offshore waters of Lake Erie. In November, hordes of migrant Bonaparte's gulls gather by the thousands, and tens of thousands of red-breasted mergansers swarm in the lake. In swirling masses so thick they resemble fast-moving storm clouds, over 100,000 mergansers have been tallied passing by within a few hours. This is also the time of year to watch for migrant loons and tundra swans. Veteran birders enjoy looking for rarities like king eider, Northern gannet, black-headed and mew gulls, and jaegers.

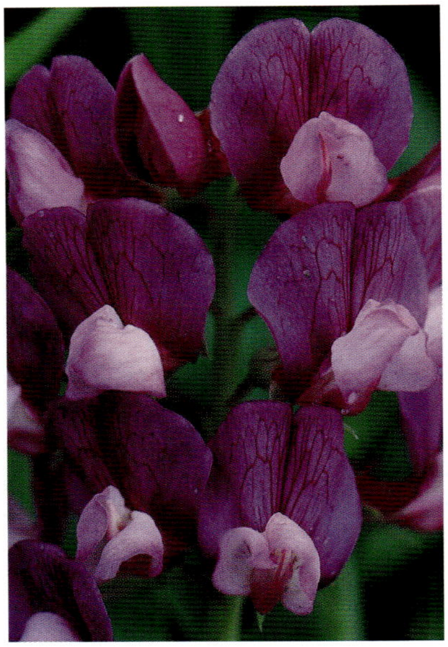

Top: The pink and purple flowers of this threatened species brighten the grassy dunes of Headlands. Inland beach pea has an amazing circumboreal distribution, occurring on beaches at northern latitudes around the globe. Its far-flung range is accounted for by the seeds, which can remain viable and buoyant in water for up to five years.

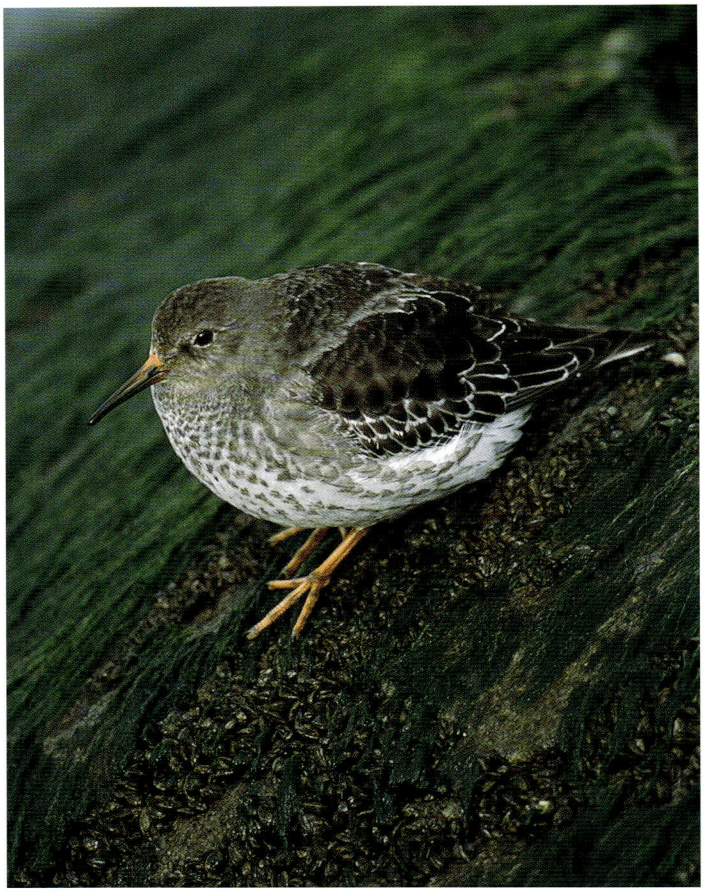

Right: Rare away from the Atlantic coast, small numbers of purple sandpipers migrate through the Great Lakes in November and December. The stone jetty at Headlands beach has become the premier locale in Ohio for finding this rarity.

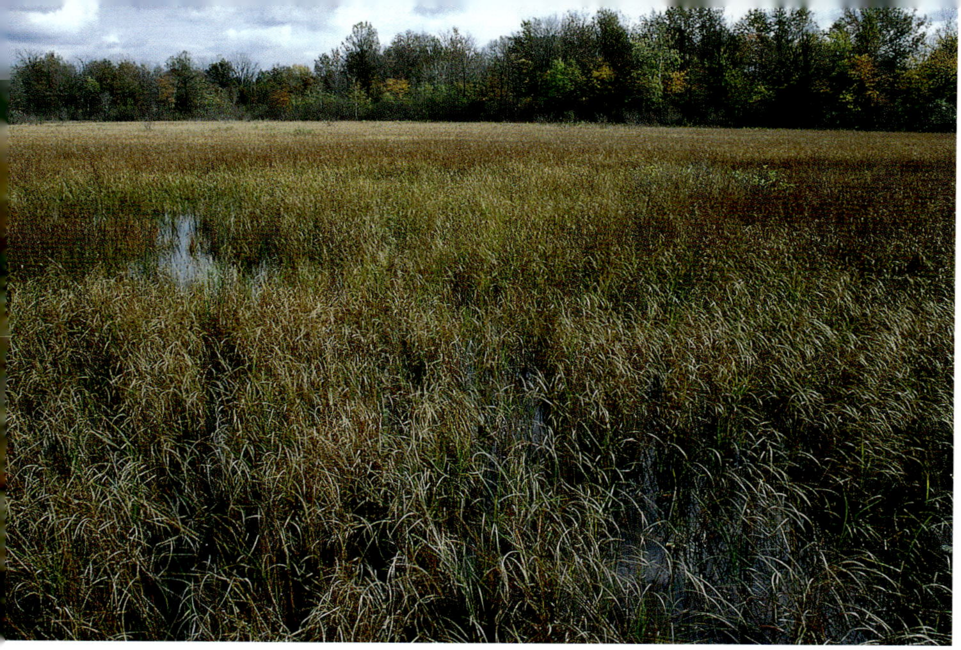

Irwin Prairie State Nature Preserve

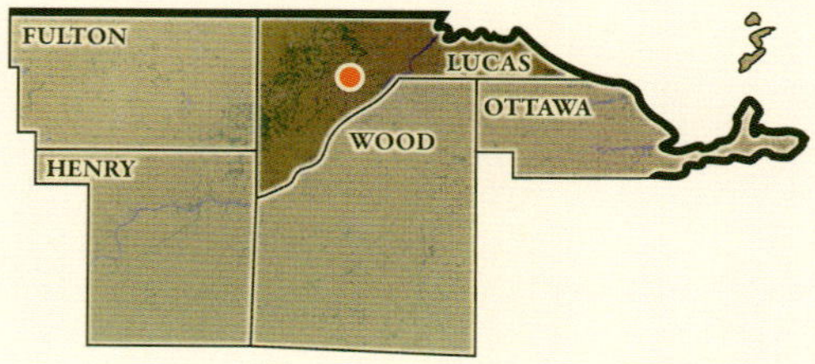

ACREAGE: 187

OWNERSHIP: Ohio Division of Natural Areas and Preserves

ACCESS: Open to the public during posted hours

WEB SITE RESOURCES: http://www.dnr.state.oh.us/dnap/location/irwin_prairie/tabid/946/Default.aspx

Irwin Prairie is Ohio's counterpart to the Everglades. The sedge that dominates the meadows is twig-rush. Its only eastern equivalent is saw-grass, which blankets the vast Florida wetlands.

The area is waterlogged year-round; an impervious clay layer underlies the wetlands, trapping water on the surface. Fortunately, a mile-long boardwalk allows easy access to the meadows, offering an opportunity to observe unusual flora and fauna. And there's plenty of that—at least twenty-six species of rare plants have been found. Many are sedges and other obscure species that require a skilled botanist to identify. Others are obvious.

Sharp eyes might spot the saclike flowers of large yellow lady's-slipper in May. Stroll to the observation tower in the wettest part of the meadows and watch for the three-parted white blooms of endangered grass-leaved arrowhead. In late fall, gorgeous blossoms of small fringed gentian can be seen along the walkway, their rich cobalt flowers looking as if they were spun from silk.

A captivating avian spectacle commences in March. Wilson's snipe, an odd sandpiper with a long bill, begin their courting antics. Males circle skyward and, at the apex of their ascent, suddenly drop landward. Wind whistling through their uniquely fluted tail feathers creates a surreal winnowing sound unlike any other aural display in Ohio.

Interesting marsh birds dwell in the wettest portions of the prairie. American and least bitterns have nested, as has Virginia rail. The metallic trills of swamp sparrows are easily heard, and sometimes marsh wrens add their sewing machine–like chatter to the mix. Rare Ohio

Top: The dominant plant in the wet meadows at Irwin Prairie is a sedge—twig-rush. Its close relative and ecological counterpart is the saw-grass of the Florida Everglades.

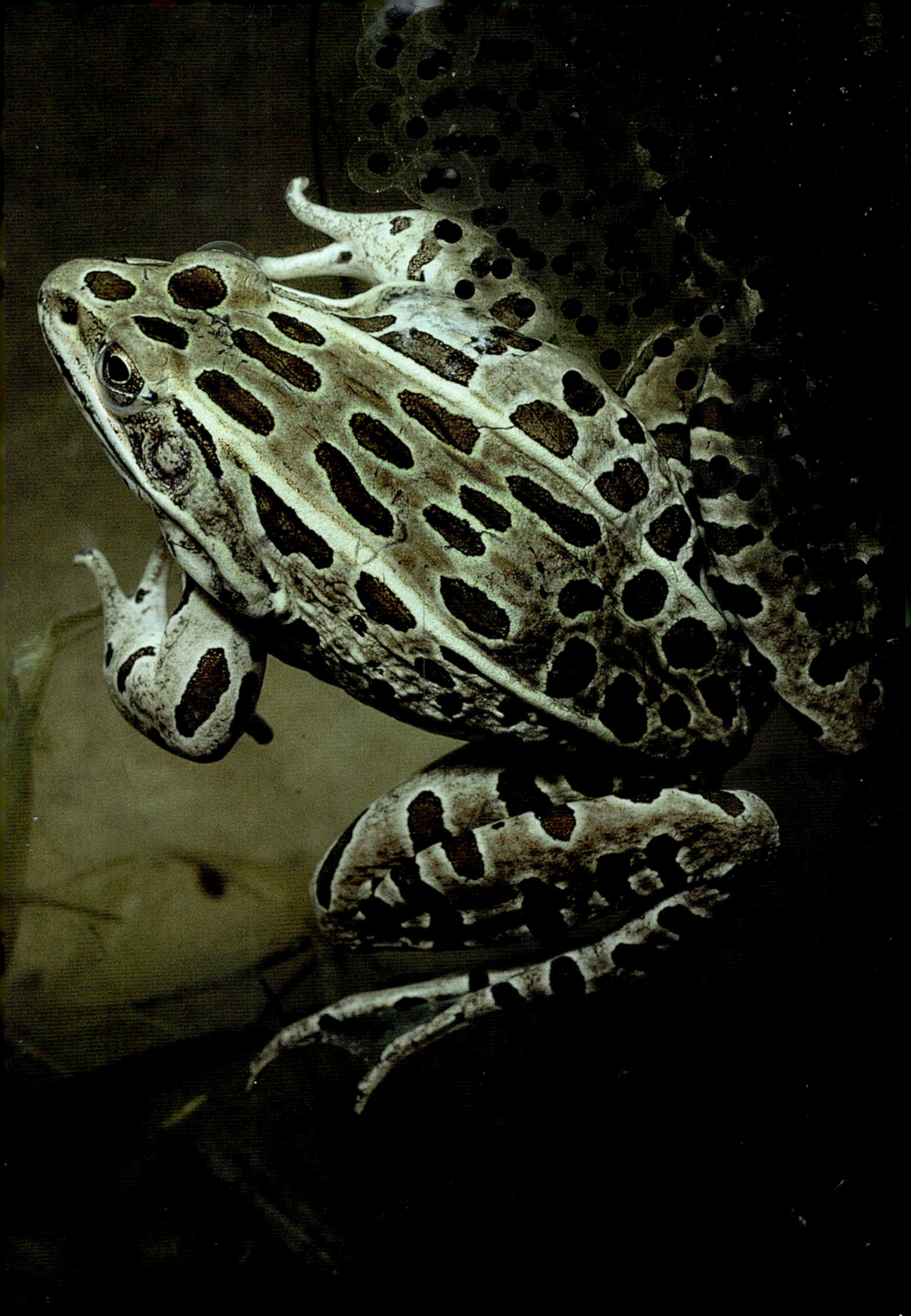

nesters such as Canada and mourning warbler have been found in June and may occasionally breed.

Irwin Prairie produces amphibians by the score. Early spring symphonies of spring peepers and western chorus frogs can be deafening, drowning out lesser songsters. Frogs are vital to wetlands; each individual is a voracious predator, consuming large quantities of mosquito larvae and other insects.

Oak Openings wetlands like Irwin Prairie are difficult to protect. As development expands in the region, insidious threats arise that impact sensitive habitats. Basements and waterline excavations wick groundwater away, speeding desiccation of nearby moist habitats. Drying of wetlands gives an edge to invasive plants like glossy buckthorn, a scourge of Oak Openings ecosystems. Extensive areas of Irwin have been overtaken by this noxious shrub. Fortunately control measures implemented by the Ohio Division of Natural Areas and Preserves have beaten it back, at least temporarily.

No place in Ohio resembles the Oak Openings, nor hosts the tremendous number of rare species found here. Lucas County, the Oak Openings epicenter, has more state-listed rare plants and animals than any other Ohio county. Conservation of these sensitive habitats is a challenge, especially as development pressures grow.

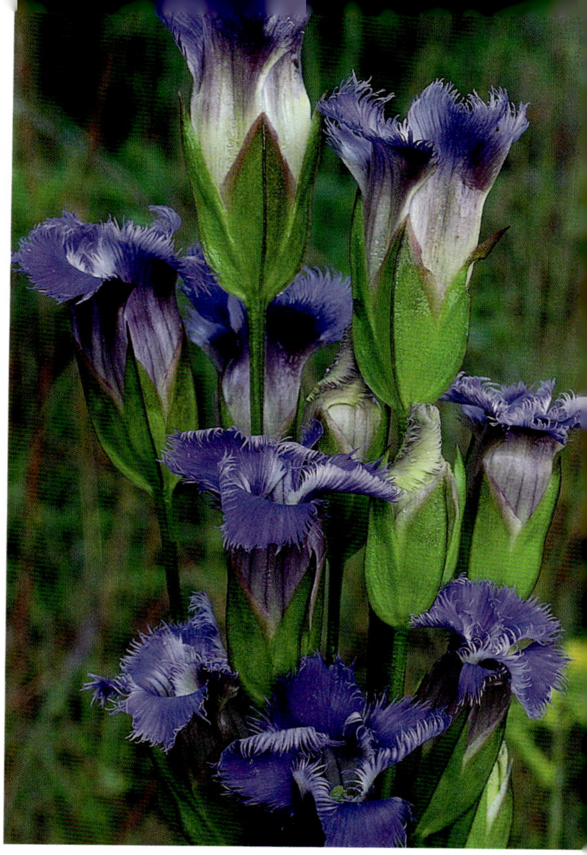

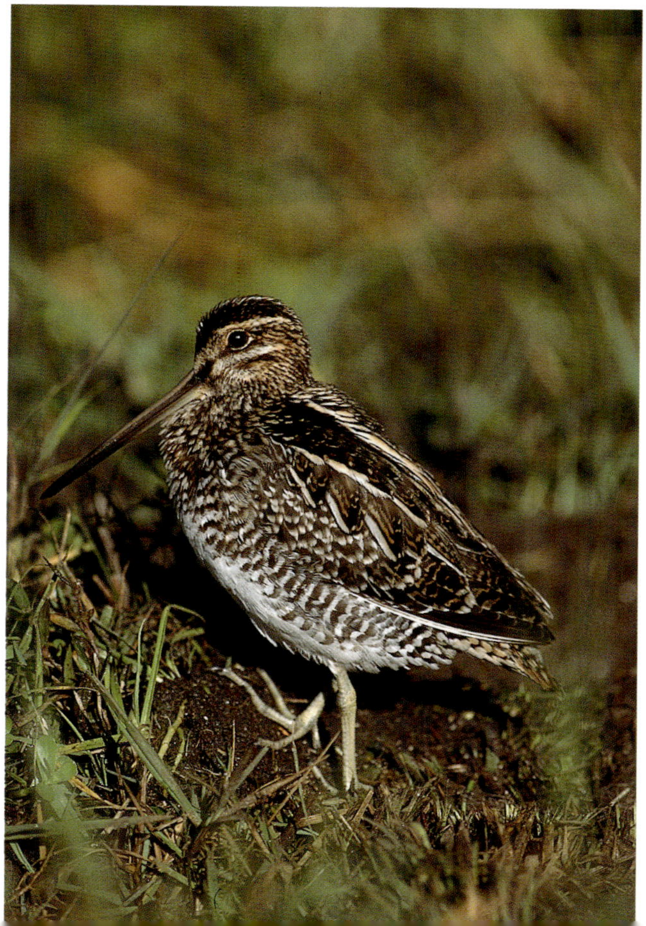

Opposite: Breeding early in the spring, Northern leopard frogs utilize vernal pools. Their young leave before the pools dry out.

Above: The small fringed gentian appears in late summer, growing in wet meadows and seeps. The beauty of this plant inspired poet William Cullen Bryant to pen a poem entitled "To the Fringed Gentian."

Left: Wilson's snipe are at the extreme southern end of their breeding range in Ohio. Irwin Prairie is one of few reliable sites where the peculiar aerial courtship displays of male snipe can be observed.

Kelleys Island

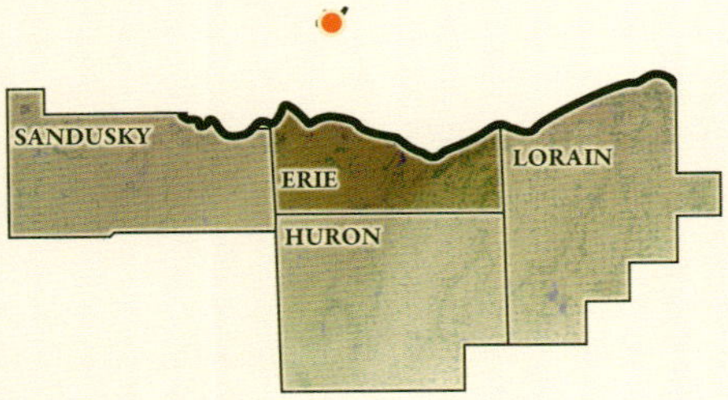

ACREAGE: 2,800

OWNERSHIP: Ohio Department of Natural Resources (710 acres), Cleveland Museum of Natural History (120 acres), the remainder private

ACCESS: Park and museum lands open to public during posted hours

WEB SITE RESOURCES:
http://www.dnr.state.oh.us/dnap/location/alvar/tabid/916/Default.aspx

http://www.dnr.state.oh.us/dnap/location/NorthPond/tabid/952/Default.aspx

http://www.dnr.state.oh.us/parks/lakeerie/tabid/753/Default.aspx

http://www.cmnh.org/site/Conservation_NaturalAreas_Map_KelleysIsland.aspx

http://www.kelleysislandnature.com/

The Lake Erie islands have been heavily developed, and most have lost much of their natural attributes. Kelleys Island, thanks to conservation efforts by the Ohio Department of Natural Resources and the Cleveland Museum of Natural History, has retained many of its ecologically significant features.

Devonian-age rock that forms Kelleys Island is part of the Columbus limestone formation. This rock formation is harder and more resistant to erosion than the surrounding strata, which have eroded away leaving elevated knobs, now surrounded by the waters of Lake Erie. These offshore islands are referred to as the Marblehead archipelago. Kelleys Island is the geological standout because of its famous grooves etched deeply in the rock, an excellent example of glacial scouring.

Much of Kelleys Island's eighteen-mile shoreline is rimmed in low, rocky cliffs. On the exposed northern exposure of the island is a superb example of an alvar, one of Ohio's rarest habitats. Alvars are wave-battered rocky shorelines that support specialized flora, and this one is protected as North Shore Alvar State Nature Preserve. Many beautiful, intriguing plants grow on the alvar. In May, golden blooms of balsam squaw-weed and deep purple Northern bog violets stand in stark contrast to their barren rocky habitat. Both plants are endangered in Ohio.

In warmer months, offshore waters sport plenty of Lake Erie water snakes. Like mini seaserpents, these rare reptiles often swim with just their heads protruding. Demonstrating an excellent example of geographic isolation, this subspecies of the widespread Northern water

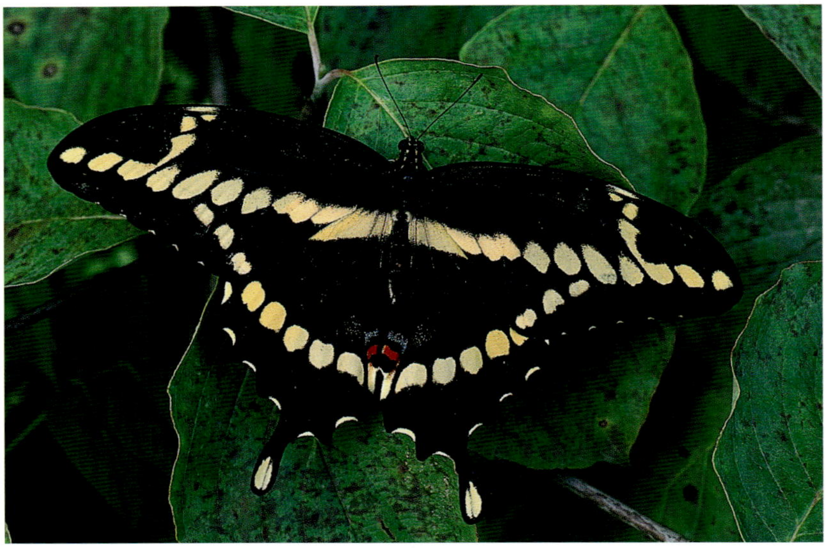

Opposite: One of Ohio's most famous and widely viewed geological features are the glacial grooves of Kellys Island. Some thirty thousand years ago, tongues of ice from the Wisconsin glacier moved south, cutting three-foot grooves in the Columbus limestone.

Right: The giant swallowtail is commonly seen on Kelleys Island, where one of its primary host plants, wafer-ash, is frequent.

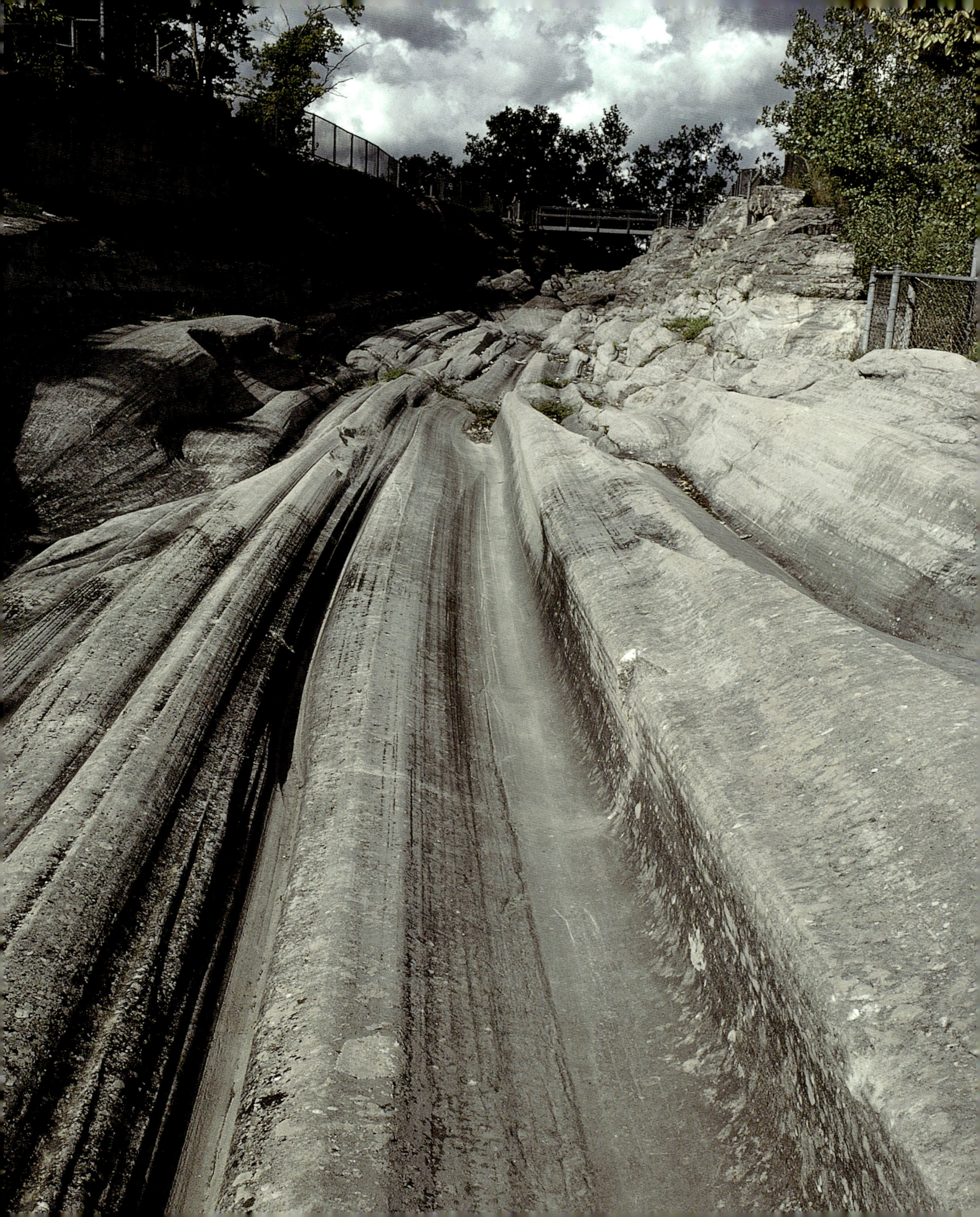

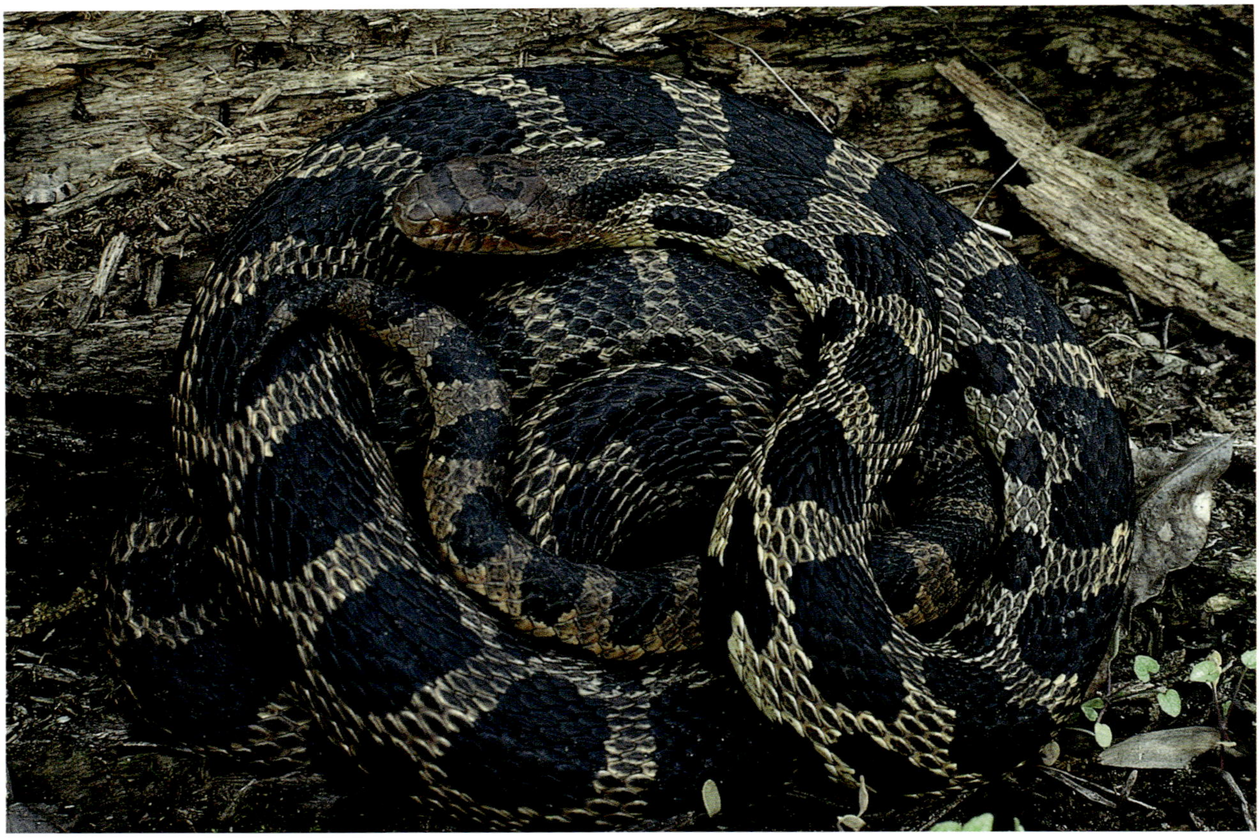

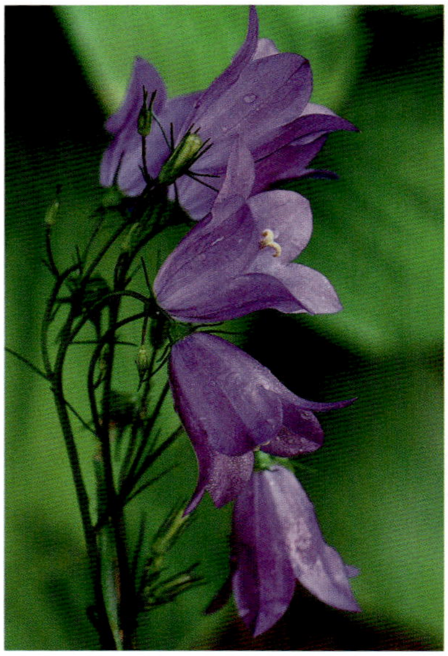

snake has become separated from the parent species and only occurs on Lake Erie islands. It is listed as federally threatened.

The finest remaining wetland on a Lake Erie island is North Pond State Nature Preserve. This mixed-emergent marsh supports native wetland vegetation like swamp rose-mallow, giant bur-reed, and the endangered wapato. Swampy woods buffer the wetland and attract scores of migrant birds. Warblers, vireos, tanagers, thrushes, and others settle here in migration to rest and refuel. Over 260 species have been recorded on Kelleys Island, including the federally endangered Kirtland's warbler. Catch a good fallout in May or September migratory peaks, and the trees can be dripping with songbirds.

Rocky shorelines and open waters of Lake Erie also harbor significant congregations of water birds. Gulls and terns of several species are common, including great black-backed gulls. This massive scavenger weighs 3½ pounds and has a 5½-foot wingspan, making it the world's largest gull. Adults are striking with charcoal-colored mantles contrasting with pure white underparts.

The smallest duck in Ohio is the thirteen-ounce bufflehead. These piebald divers are hardy, and migrant flocks collect in icy Lake Erie waters off Kelleys Island in November and December. Several thousand of the ducks sometimes gather—a significant percentage of their global population.

Late summer and fall bring migrating Monarch butterflies to the island. These globetrotters are powerful fliers and routinely cross Lake Erie. Compounds within the butterflies engage the earth's magnetic field, pulling them on a steady southwestern course toward Mexican wintering grounds. Kelleys Island offers a resting spot after the lake passage, and occasionally masses of butterflies encrust the trees.

Top Left: The Eastern fox snake has one of the most localized distributions of any North American snake, occurring only in Michigan, Ohio, and Ontario, Canada. In Ohio, they are only found on the Lake Erie islands and in some northwestern counties.

Left: Harebells typically grow on wave-splashed limestone cliff faces above the waters of Lake Erie. Endangered in Ohio, harebells only occur on the Lake Erie islands, with the exception of one isolated Miami County population.

Right: One of only six federally listed rare plants in Ohio, the state-endangered Lakeside daisy is only known from the Marblehead Peninsula of Ohio, and a few sites in Michigan and Ontario, Canada. While not indigenous to Kelleys Island, plants were transplanted here from Marblehead—three and one-half miles away—in the early 1990s to establish new colonies. The mass blooming of this very rare plant is one of Ohio's greatest botanical spectacles.

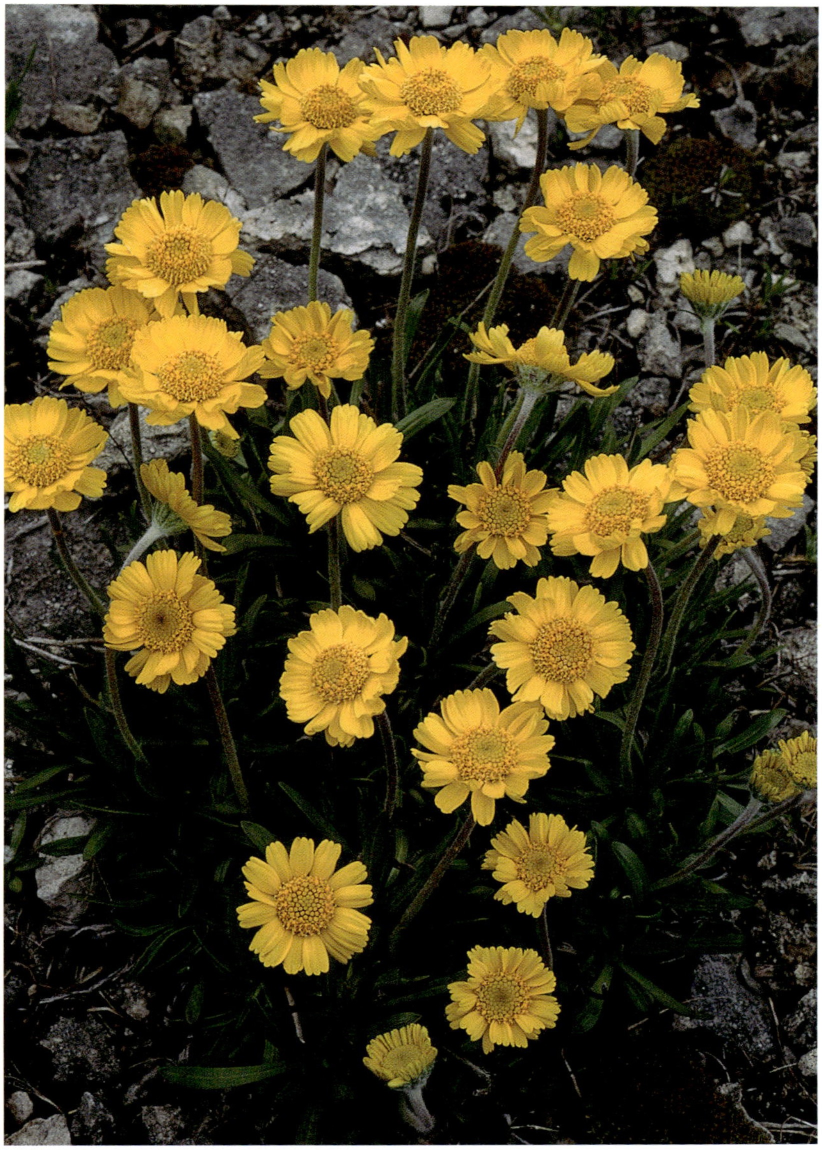

Kitty Todd Preserve

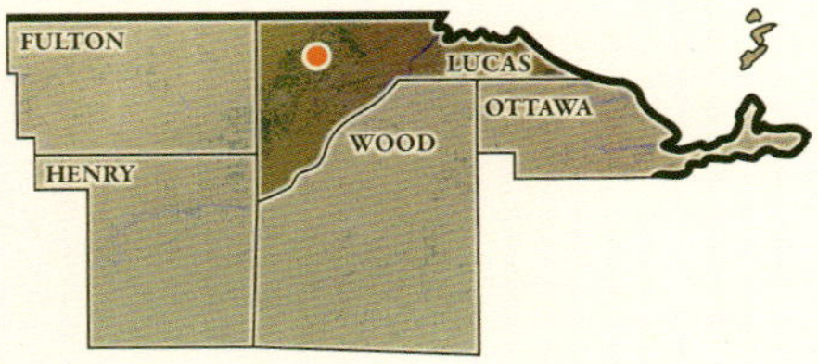

ACREAGE: 500

OWNERSHIP: The Nature Conservancy

ACCESS: Open to the public during posted hours

WEB SITE RESOURCES:
http://www.nature.org/wherewework/northamerica/states/ohio/preserves/art162.html

http://www.dnr.state.oh.us/tabid/949/default.aspx

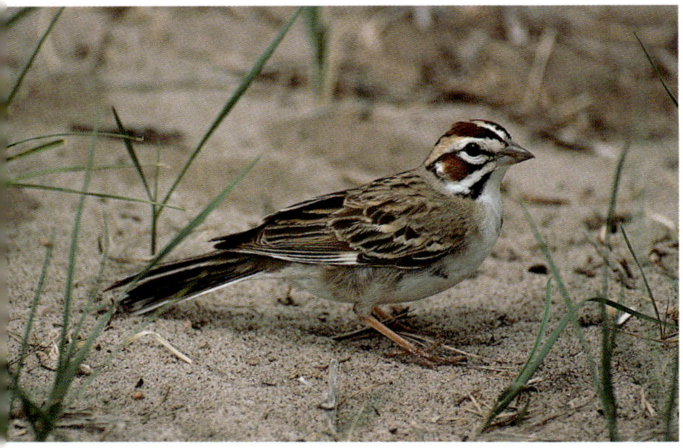

In a good year, there probably aren't more than a few dozen pairs of lark sparrows breeding in Ohio. Shifting eastward from the Great Plains during the warmer and drier Xerothermic period, the Oak Openings population is the easternmost known.

The Oak Openings were named for the open, parklike savannas that were present at the time of European settlement. Scattered large black and scarlet oaks shadowed an understory of grasses and unusual herbaceous flora—a landscape so unfettered by trees and shrubs that settlers could easily pilot wagons through the region.

Fire was the instrument that maintained these habitats, and Indians were the main incendiary agents. For thousands of years native tribes lit infernos that raced through the Oak Openings, torching saplings and brush but sparing thick-barked fire-resistant big oaks. Many herbaceous prairie plants co-evolved with fire, as heat and flame are essential ingredients for their survival.

Fire performs several essential roles in the Oak Openings. Burning creates open, sunny conditions necessary for sun-loving plants. Periodic blazes scrub duff layers—dead mats of plants—from the soil, promoting easier growth of new plants. Following a burn, massive infusions of ash coat the soil, providing a burst of nutrients into the ground. Lastly, some prairie plants seem to require the tremendous heat of burns to crack dense seed coats and permit germination.

Man fears seemingly uncontrolled wildfires, and for roughly two hundred years postsettlement, fire was largely suppressed. After The Nature Conservancy acquired Kitty Todd, managers began to reintroduce fire. The results have been spectacular, and this preserve is among the finest examples of an Oak Openings savanna ecosystem.

In May, clumps of blue-flowered wild lupine dot sandy savanna soils. This plant nearly disappeared during the fire suppression era as its open haunts became overgrown with invading trees and shrubs. Tangential to lupine decline was the loss of the Karner blue butterfly. Now that the butterfly's host plant has made a comeback, efforts to reintroduce the blue to Kitty Todd have met with some success. This is currently the only Ohio locale where enthusiasts have a shot at finding this endangered species.

A large snake of the sandy barrens is the blue racer. They can exceed six feet in length and are tinted in rich indigo hues. As the name implies, they are speedy, and adults can hit twelve miles per hour for short bursts. Hunting racers move about with heads elevated, cobra style, and look intimidating. They're best left alone; if molested, racers won't hesitate to bite. Racers are arboreal and may climb trees when pursued.

Damp swales concentrate rare wetland plants, many of which seldom or never occur outside the Oak Openings in Ohio. In early July, white spires of colic-root flowers pepper low-lying areas of the

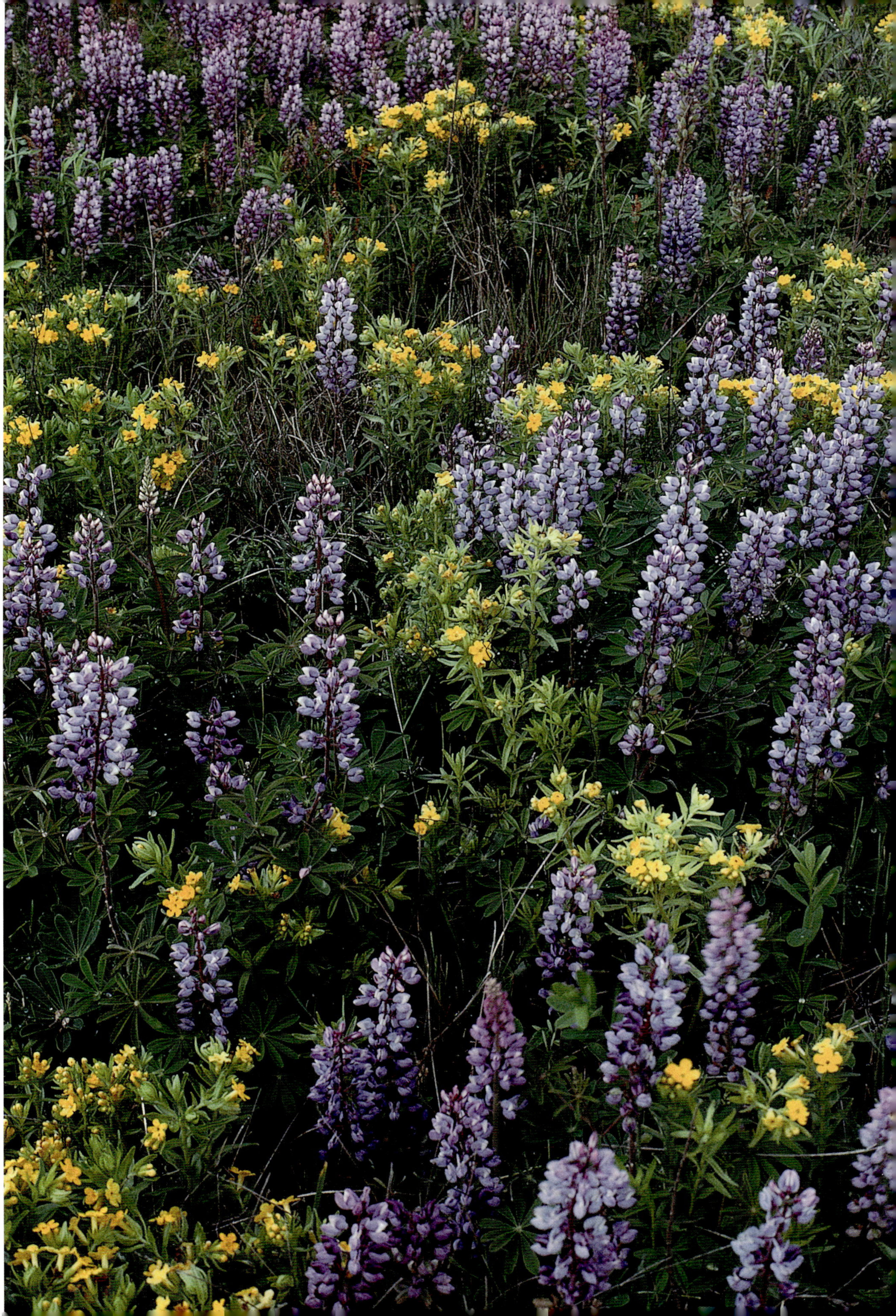

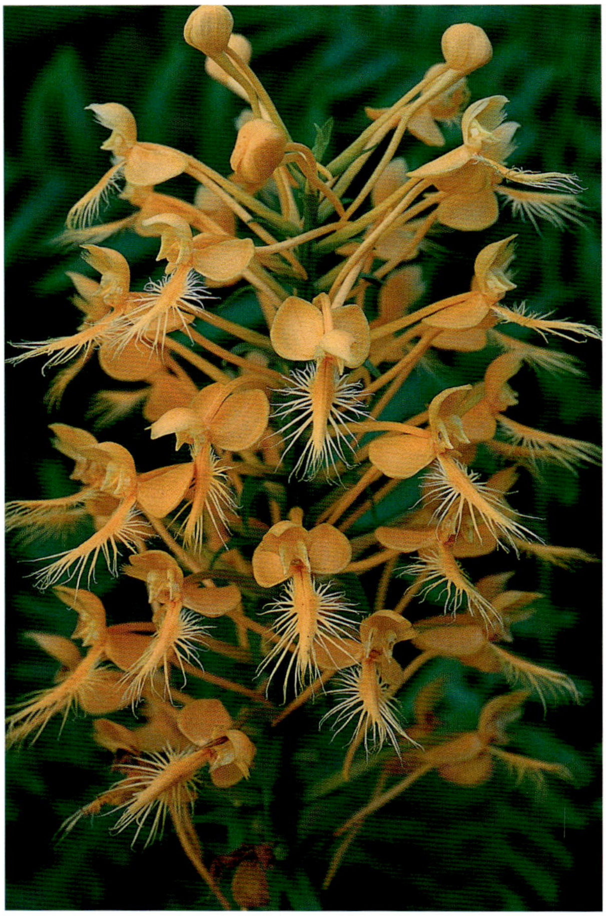

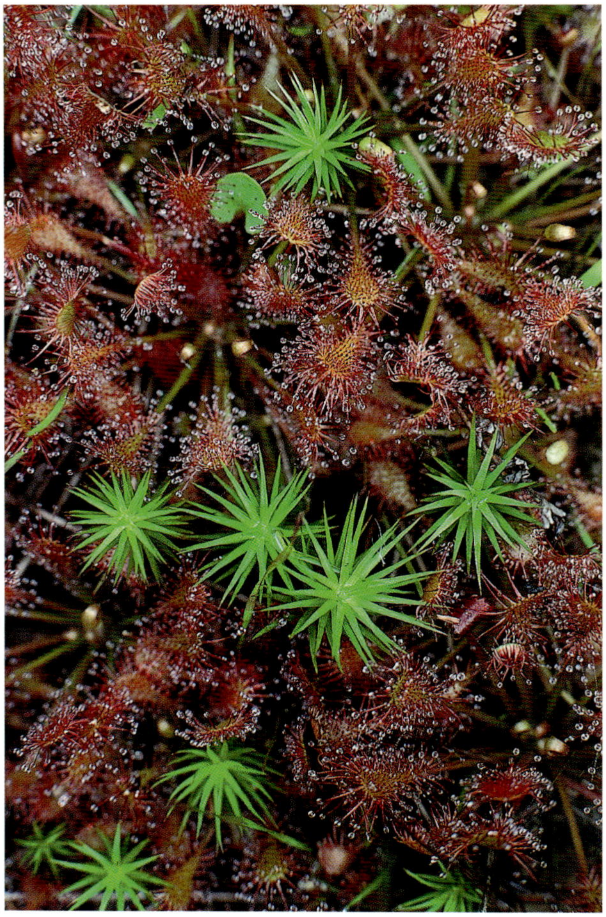

Overleaf: Wild lupine, with its racemes of blue flowers, contrasts with the orange flowers of plains puccoon. Both plants are rare in Ohio. Wild lupine is the host plant for the endangered Karner blue butterfly.

Top left: Forty-six native orchid species have been recorded in Ohio. Like the threatened yellow fringed orchid, most have become rare due to habitat loss.

Top right: Almost an Oak Openings exclusive, the spathulate-leaved sundew grows on moist sand. Sticky hairs on the leaves trap insects.

prairie. This rare lily produces an interesting aloelike rosette of broad leaves. Where spikes of colic-root blooms are seen, further botanical investigation is warranted. Other rarities often accompany it, such as twisted-yellow-eyed-grass, Atlantic blue-eyed-grass, cross-leaved milkwort, and lance-leaved violet.

The highest, driest sand knolls may shock botanical neophytes. Some are carpeted with thorny growths of prickly-pear, Ohio's only native cactus. They bear stunning yellow blossoms in June. Other rare xerophytes grow nearby, including Bicknell's sedge, scaly blazing-star, and dotted horsemint. Among the rarest is the endangered prairie fern-leaved false foxglove. Despite its cumbersome moniker, it is one of our showiest wildflowers and is known from only a handful of Oak Openings locales.

Ohio's greatest natural resource and most dynamic geological feature is Lake Erie. Only sixteen thousand years old, this geological youngster has been in its present configuration for only five thousand years. Shoreline conditions change rapidly, and beaches and wetlands have shifted dramatically over the past century. Events called seiches, caused by strong winds, can pile up water sixteen feet higher than on the downwind side of the lake. Overnight, wetlands with two feet of water become mudflats, only to fill again later that day.

While natural phenomena has changed western Lake Erie's vast marshes, so have people in the past 150 years. Much former marshland has been drained and converted to farmland. Most areas have been impacted by dike construction to manage water levels. Roads and railroad tracks have obstructed natural drainage. Nonetheless, the western Lake Erie marshes remain our biggest contiguous wetlands and one of the most significant biological hotspots in North America.

Magee Marsh is one of the most famous birding locales in the country. Tremendous fallouts of neotropical migrants fill trees and bushes in May, attracting thousands of observers from all over the country. On International Migratory Bird Day (the second Saturday in May), over five thousand birders often converge here. They come to visit a seven-acre woodlot bisected by the "bird trail." This elevated boardwalk traverses wet woods of box-elder, cottonwood, and green ash, and migrant birds are often at one's fingertips.

If suitable weather conditions create a "big day," this coppice of trees may be packed with migrating songbirds. Mid-May brings

Magee Marsh Wildlife Area/ Ottawa National Wildlife Refuge

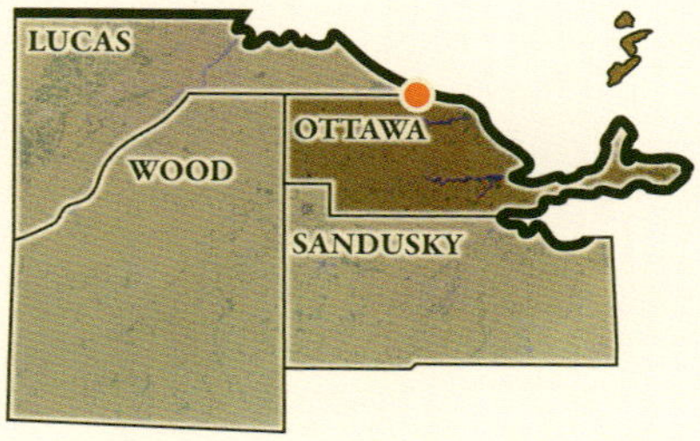

ACREAGE: 11,000 (Magee 2,000; Ottawa 9,000)

OWNERSHIP: Ohio Division of Wildlife, U.S. Fish & Wildlife Service

ACCESS: Open to the public during posted hours. Some areas off-limits

WEB SITE RESOURCES:
http://www.dnr.state.oh.us/Portals/9/Images/wildarea/magee.gif

http://www.ohiobirds.org/birdingsites/showsite.php?Site_ID=69

http://www.fws.gov/midwest/Ottawa/

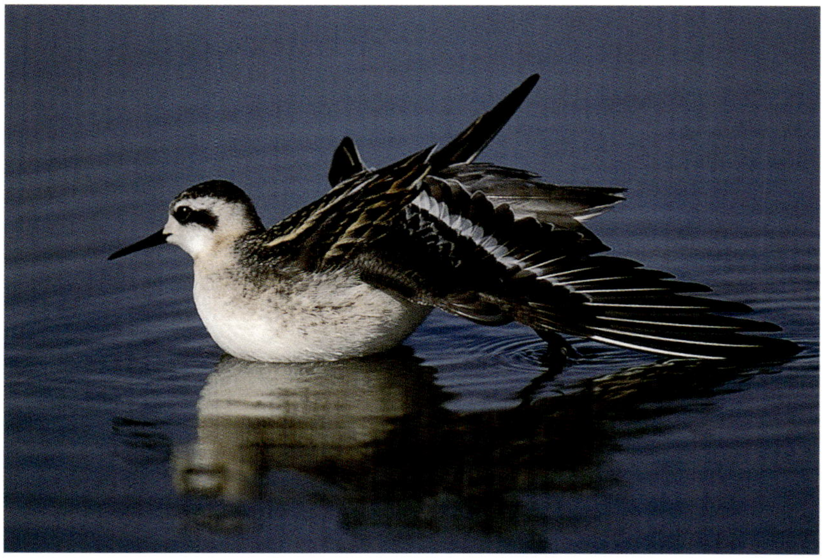

Small numbers of red-necked phalaropes pass through Ohio en route between Arctic breeding sites and marine wintering grounds in southern oceans. The impoundments at Ottawa can be a good place to view shorebirds.

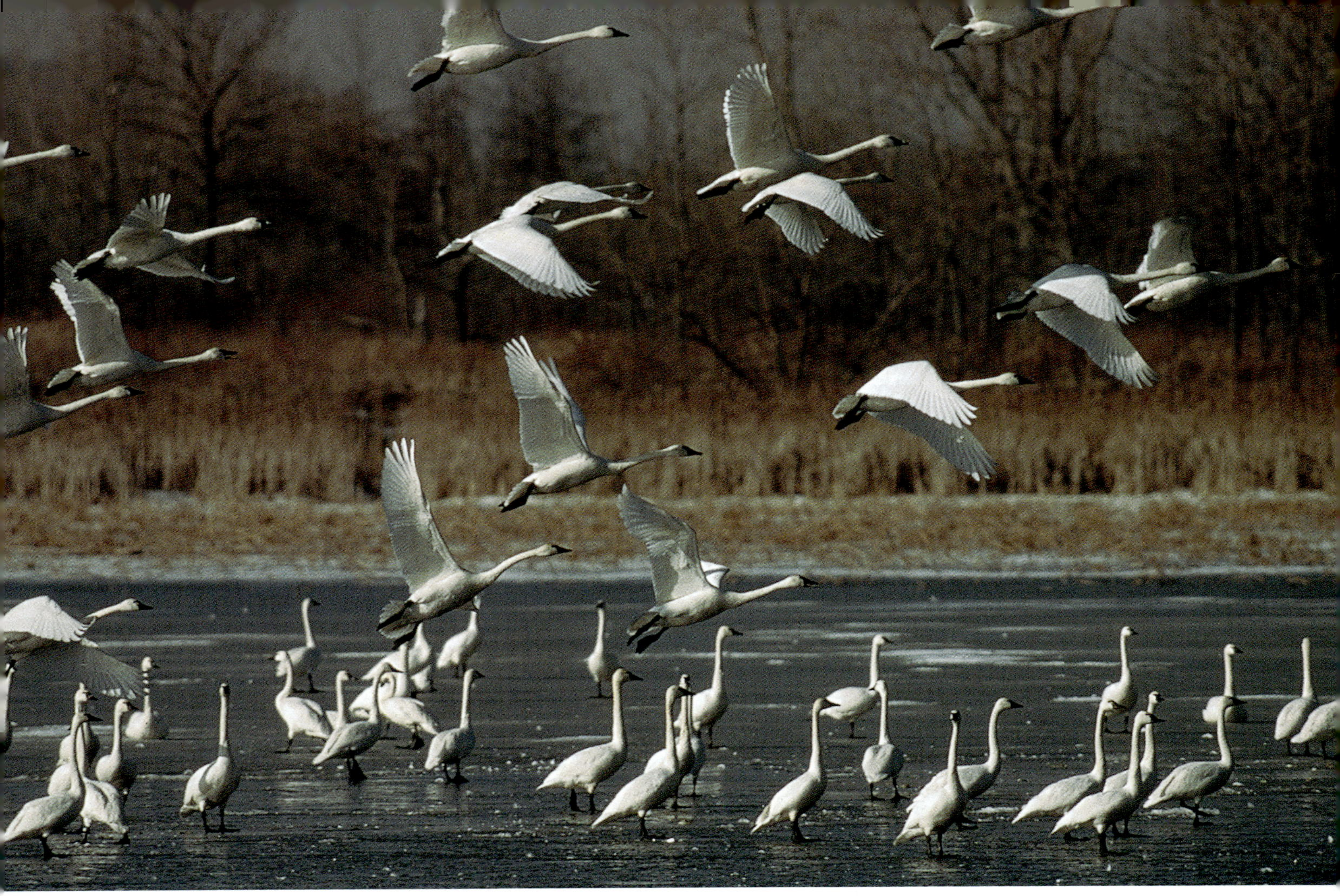

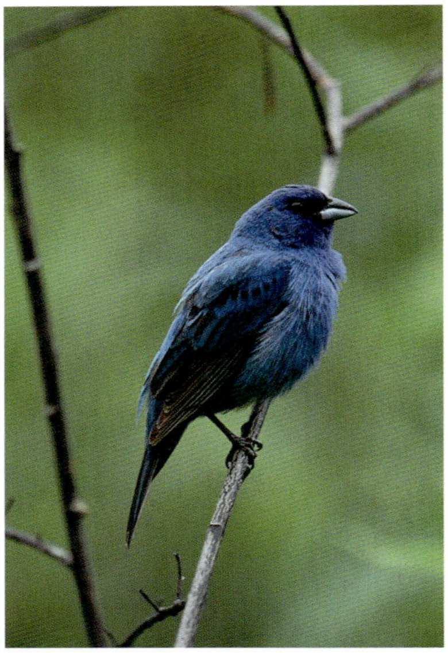

peak numbers and diversity, and a skilled birder might tally over one hundred species in a morning. Staggering numbers of colorful jewels like scarlet tanager, Baltimore oriole, rose-breasted grosbeak, and indigo bunting are everywhere, with warblers generating the most excitement. All thirty-seven species that migrate through Ohio pass through this region, and most can be seen on a good day at Magee. Scores of magnolia, black-throated green, Nashville, bay-breasted, and blackpoll warblers feed in a frenzy along the trail. This is the best spot in Ohio to catch the federally endangered Kirtland's warbler. More migrants have been detected at Magee and vicinity than any other locale.

To reach the bird trail one must drive the causeway road that bisects one of Ohio's finest marshes. Filled with diverse native flora, these wetlands support myriad migrant and breeding birds. All of the regularly occurring waterfowl can be found, as well as rarer birds such as king rail, American and least bitterns, black tern, and snowy egret.

The overall feel is Floridian, with dozens of great egrets and great blue herons stalking prey in roadside ditches.

Just to the west of Magee is Ottawa National Wildlife Refuge, Ohio's only federal refuge. Part of Ottawa includes rocky West Sister Island, an eighty-two-acre outcrop nine miles out in Lake Erie. Our only federally designated wilderness area, the island supports a rookery of several thousand water birds, mostly great blue herons, black-crowned night-herons, great egrets, and double-crested cormorants. Most herons seen foraging at Magee and Ottawa nest on West Sister Island.

Top Left: March and November are prime months to find migrant tundra swans at Magee Marsh and vicinity. During mild winters, large numbers remain in the region.

Left: Upon their return from wintering grounds in the Caribbean and in Central America, large numbers of indigo buntings can bunch up at migratory hotspots like Magee Marsh. This is a common breeding species, found in all eighty-eight counties.

Right: The year 1979 marked the low point of breeding Ohio bald eagles, with only four nests. With the ban on DDT, eagle populations rebounded. In 2008, there were 184 nests known in Ohio.

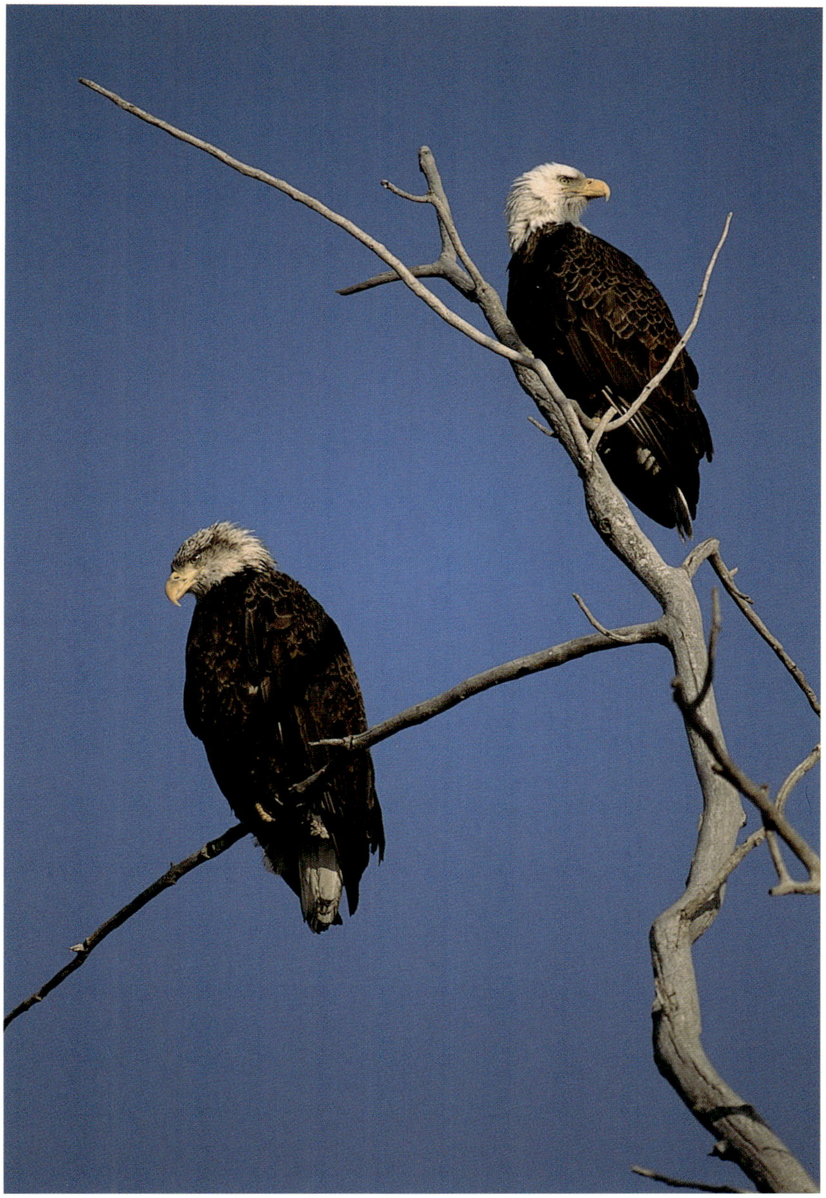

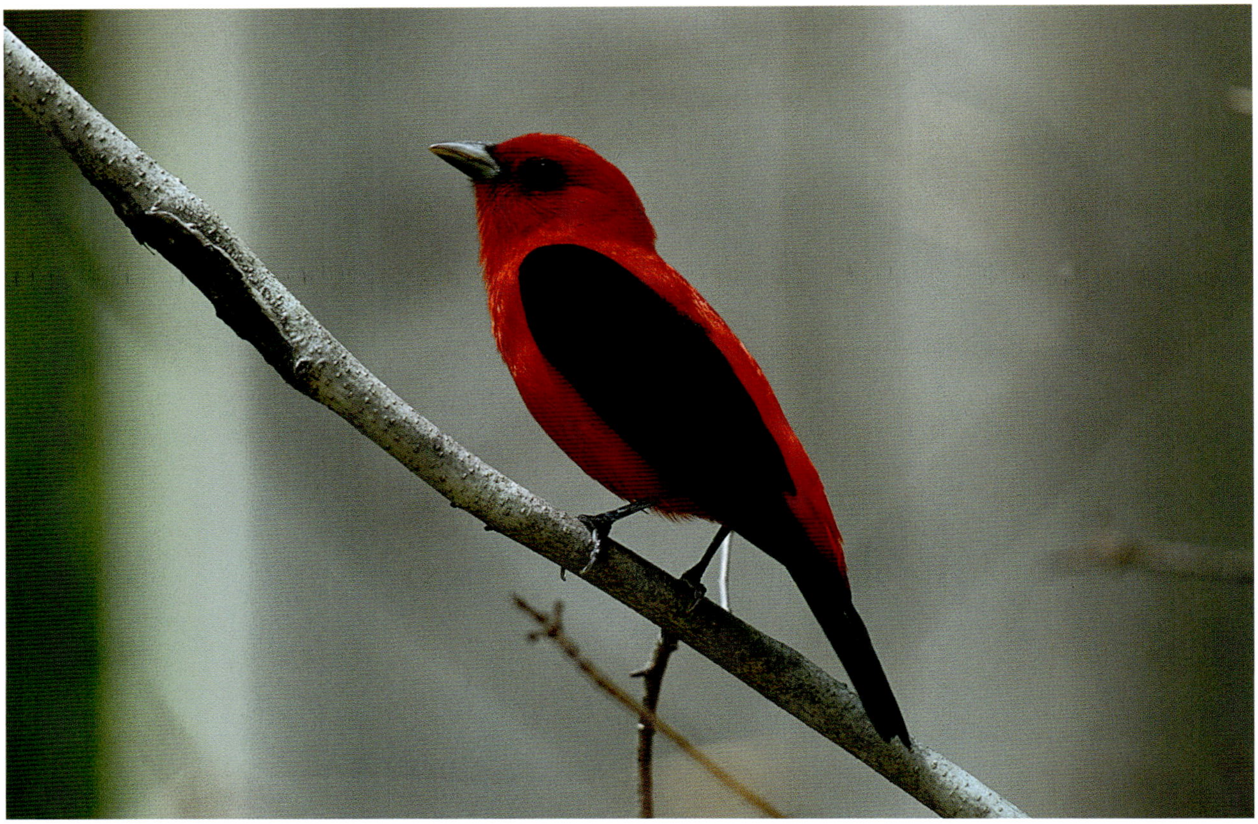

The raspy cadences of scarlet tanagers are a common sound in Ohio woodlands. One can get great looks at this species as well as many other neotropical migrants at the Magee Marsh boardwalk.

Ottawa's several thousand acres of marshes lure birds galore, and when low water levels coincide with shorebird migration, scores of waders congregate. Most plovers and sandpipers breed in the high arctic and spend the winter along the southern U.S. coast and as far south as Argentina. Resting habitat is vital in transit, and Lake Erie marshes are critical stopovers for these transglobal wanderers. August mudflats can support up to twenty-five species simultaneously, including rarities like marbled godwit, American avocet, and buff-breasted sandpiper.

Stubby little semipalmated sandpipers, which stop at Ottawa in large numbers, illustrate the global significance of this resource. These six-inch shorebirds weigh less than an ounce and breed in tundra 1,200 miles or more to the north of Ohio. They winter along South American coasts, and some of these diminutive sprites travel 8,500 miles one way, 17,000 miles round-trip. Vast mudflats that allow them to rest and refuel are essential. How we protect and manage Ohio lands links intimately to global biodiversity.

No area in Ohio supports a greater density of rare flora and fauna than this 130-square-mile region. At present, over one hundred species of state-listed rare plants and at least twenty rare animals have been documented.

Sand defines the Oak Openings; the region is like an enormous sandbox. To what do we owe all this sand? Lake Erie, from a time long ago, when it was much larger. Prior to the last glaciation about twelve thousand years ago, our Great Lake sprawled into present-day Fulton, Henry, Lucas, and Wood counties. As glacial ice retreated northward, outlets were opened that partially drained Lake Warren—Erie's much larger predecessor—stranding its beaches and dunes far from the water.

A striking example of an Oak Openings dune is near the junction of Girdham and Monclova roads. The open sandscape supports all manner of rare flora and fauna, from miniscule antenna-waving wasps to robust blunt-leaved milkweeds. Nature abhors a vacuum, though, and open sands are constantly under siege by pioneering plants. Most dunelike Oak Openings habitats have long been overgrown in a process known as ecological succession. Also, many of the wet prairies have been tiled and drained. The greatest botanical diversity occurred in the 1920s and prior, before development pressure claimed many of the wetlands.

Botanical pioneers of open sand are themselves rare. Starved panic grass and low sand sedge form dense tussocks, stabilizing shifting

Oak Openings Preserve

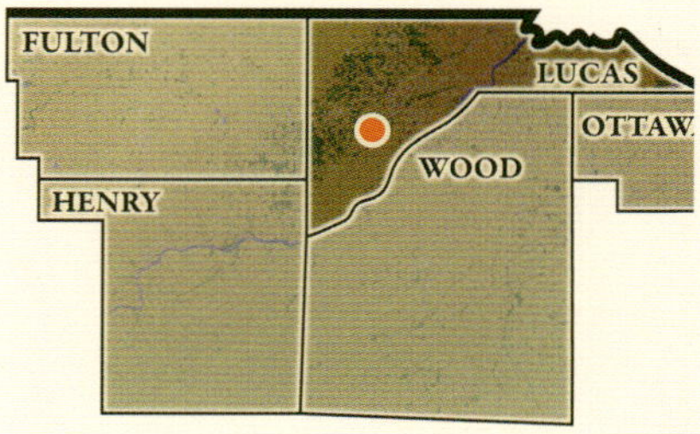

ACREAGE: 3,668

OWNERSHIP: Metroparks of the Toledo Area

ACCESS: Open to the public during posted hours

WEB SITE RESOURCES: http://www.metroparkstoledo.com/metroparks/oakopenings/

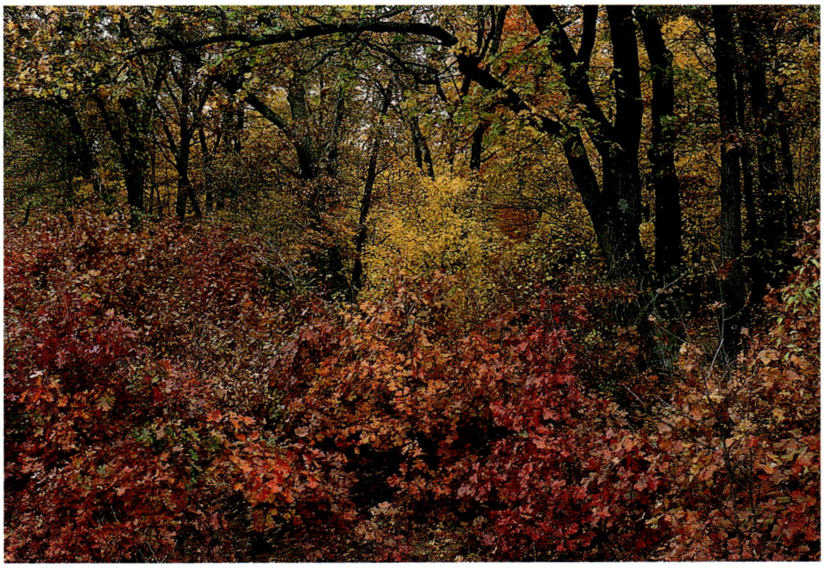

Oaks tinge woodlands in muted reds, oranges, and yellows in October. Progressive, interventive management by a variety of land-managing agencies has restored some of the Oak Openings savannas to their original condition.

sands and providing other plants a foothold. Eventually, various herbs spring up within grass and sedge colonies, followed by shrubs such as sand cherry and ultimately seedling oak trees. The ecological finale is one of mature oaks, a forest entirely different floristically than the ancestral sandy dunes.

Rare animals and rare plant communities go together, and some extraordinary beasts roam the Oak Openings. This is one of few Ohio locales where American badgers are regular. Big and ill-tempered, these burly brutes excel at burrowing. They probably migrated into Ohio four thousand years ago during the warmer Xerothermic period, hot on the heels of favored prey, thirteen-lined ground squirrels.

At the other end of the scale is the tiny Karner blue butterfly. Endangered and only found in the Oak Openings, this insect tells an ecological tale. Most butterflies are linked to specific plants, and Karner blues require wild lupine. Lupines need open sandy habitats that are maintained by fire. When fire is suppressed, as it was through much of the twentieth century, lupines disappear as shrubs and trees invade.

Below: State-endangered fringed polygala is an Oak Openings rarity. Ironically, it is an abundant species of the north woods.

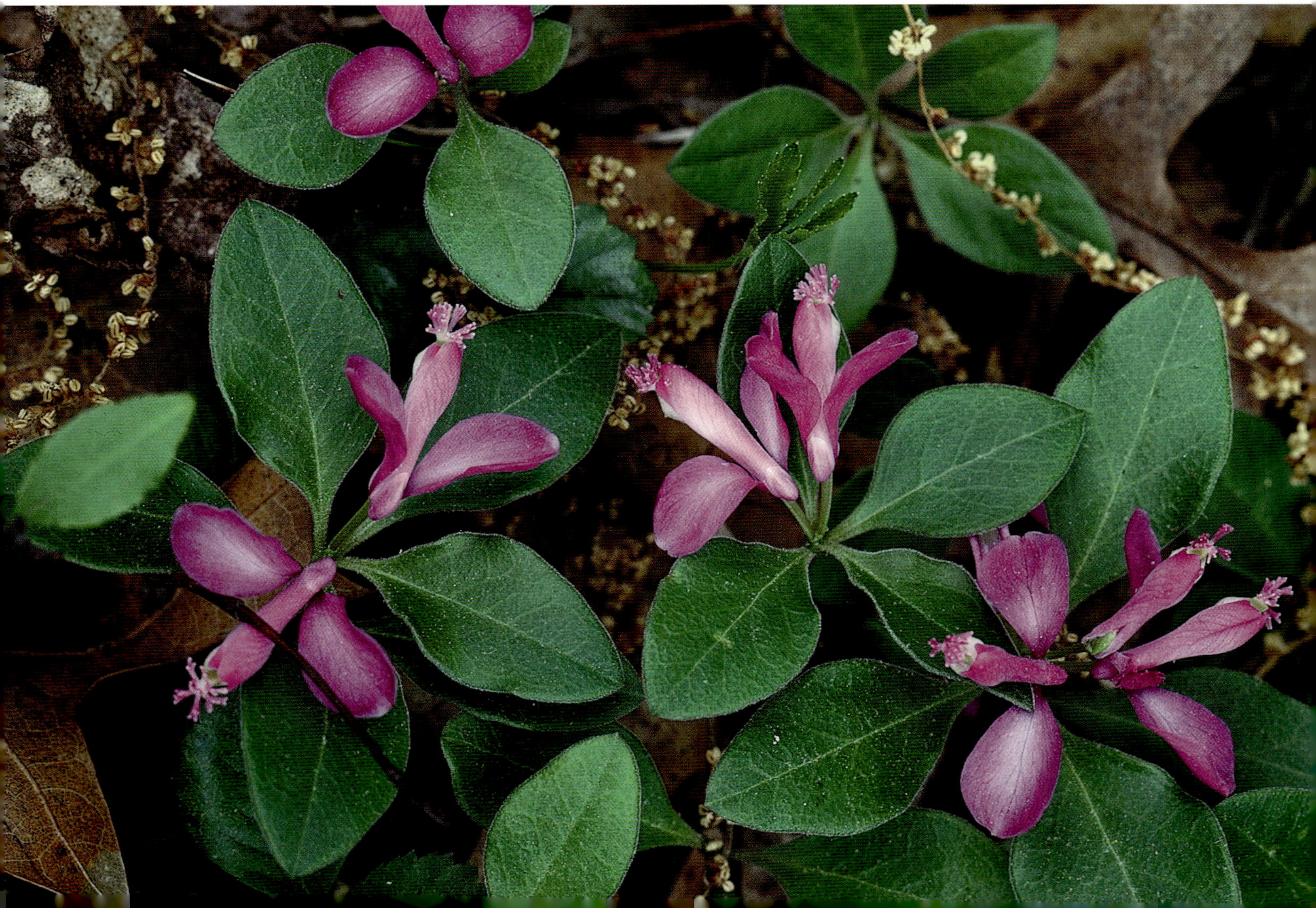

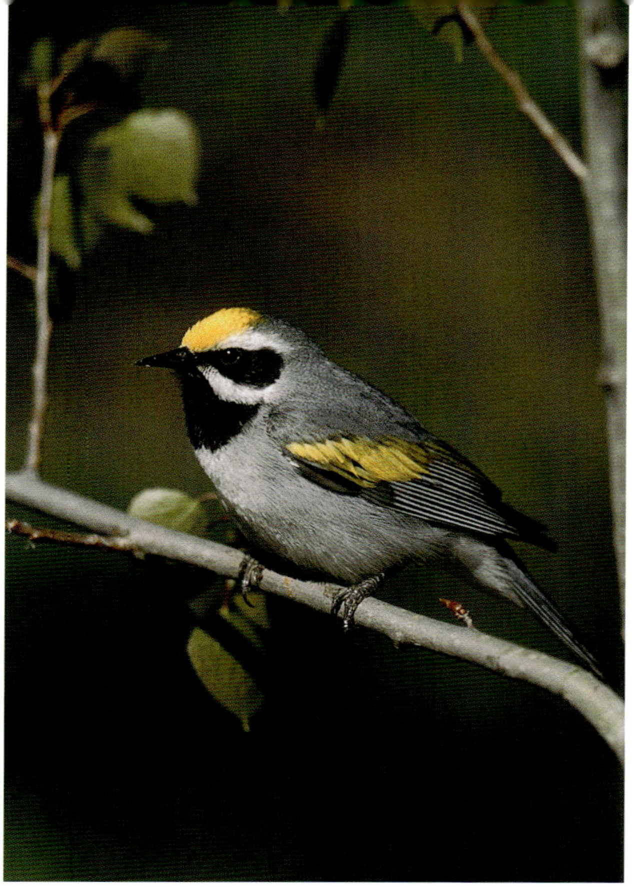

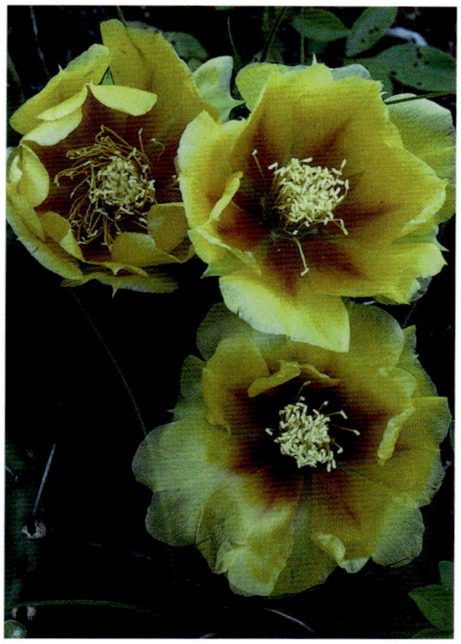

Above left: Always a very rare Ohio breeder, golden-winged warblers have essentially disappeared. Until the mid-1900s, the Oak Openings was our stronghold for them. A cause of decline is genetic introgression with a closely related species, the blue-winged warbler.

Above: Not often thought of as home to cacti, Ohio does have one native species. Prickly-pear is a relict of a time when Ohio was much hotter and drier but persists on Oak Openings dunes.

Overleaf: Eastern hognose snakes are at home in the sandy soils of the Oak Openings. This toad-eating species is a dramatic bluffer; hissing, striking, and finally playing dead if the former strategies fail to intimidate.

Several management agencies have reintroduced fire management, and lupines and Karner blues once again brighten the landscape.

One of the rarest breeding bird species in the Oak Openings is the lark sparrow. Perhaps two dozen pairs of this handsome sparrow occur in a good year. Lark sparrows also prosper by changes wrought by controlled burning. Southern specialty birds at their northern limits are also of interest. Resplendent blue grosbeaks decked in hues of rich indigo set off by cinnamon wing bars are found no farther north. Another southerner is the summer tanager. This species needs open oak woodlands, and management practices have benefited this brick-red beauty.

Young but complex, Oak Openings habitats are always interesting to explore and offer finds that can't be made elsewhere in Ohio.

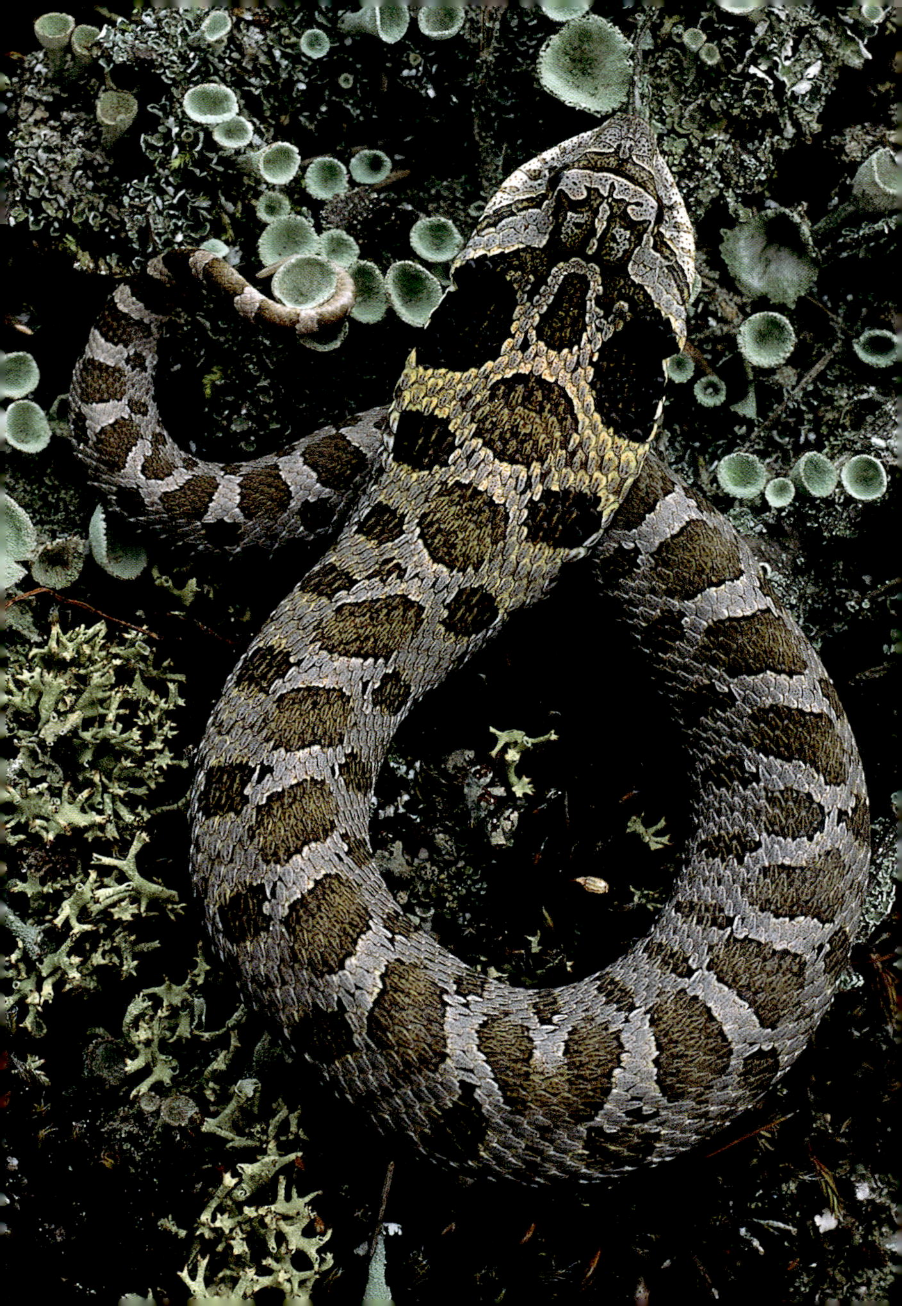

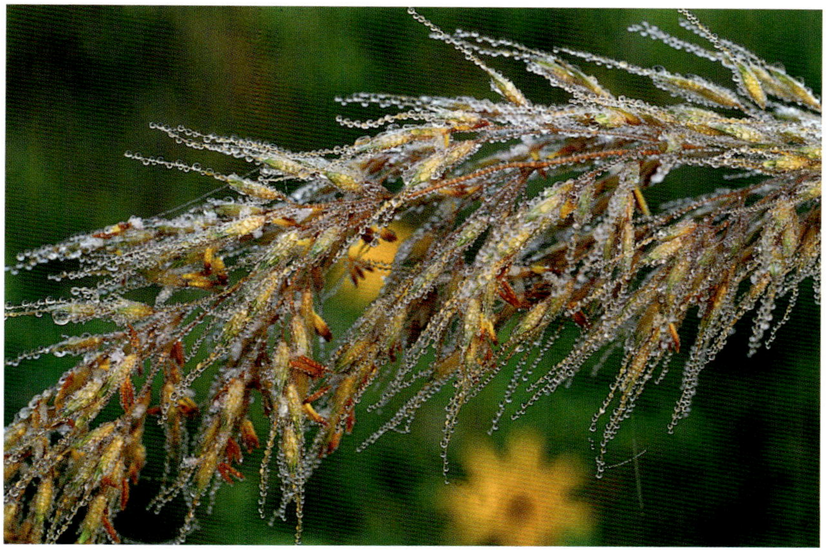

Before settlement, Castalia Prairie stretched nearly to Sandusky, merging with vast prairie wetlands buffering Lake Erie's western basin. The only significant surviving remnants are in Resthaven Wildlife Area. Prior to acquisition by the Ohio Division of Wildlife in 1942, the prairie was mined for marl, a limey mineral formed by artesian waters and used in the cement-making business. Evidence of marl extraction can still be seen in the form of linear excavations separated by elevated berms.

In spite of past disturbances, prairie plants abound. Species in the sunflower family are important prairie components, and many of them are obvious in summer and fall. Gray-headed coneflower, prairie-dock, whorled rosinweed, and black-eyed Susan all shoot forth yellow blooms that attract armies of insects. Of Ohio's thirteen species of milkweeds, eight can be found here. Preeminent is Sullivant's milkweed, which was discovered in Ohio and named for its finder, William Sullivant of Columbus. Like most milkweeds, its sweet nectar lures scores of butterflies.

There are some must-see botanical events in Ohio. The mid-May blooming of thousands of white lady's-slippers in Castalia Prairie is one of them. The Ohio Division of Wildlife manages this endangered orchid by regularly burning the meadows in which they occur. Like many prairie plants, white lady's-slippers orchids co-evolved with fire and need regular burns to survive and thrive.

Brushy wetlands and open meadows of Resthaven Wildlife Area support many species of birds, both breeders and migrants. In March,

Resthaven Wildlife Area

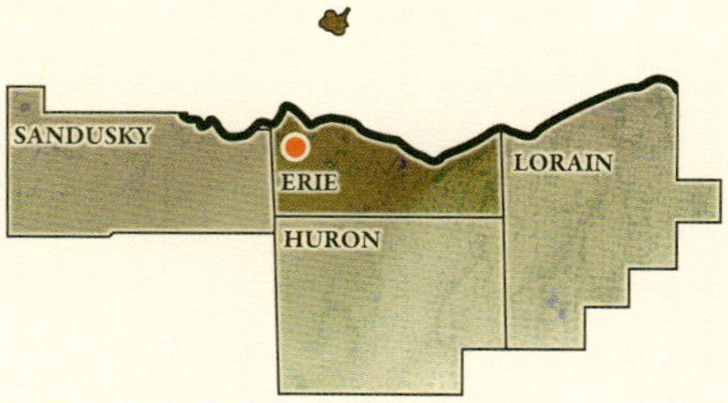

ACREAGE: 2,272

OWNERSHIP: Ohio Division of Wildlife

ACCESS: Open to the public during posted hours

WEB SITE RESOURCES:
http://www.ohiodnr.com/Home/wild_resourcessubhomepage/WildlifeAreaMapsRepository/tabid/10579/Default.aspx

http://www.ohiobirds.org/birding-sites/showsite.php?Site_ID=33

Top: Indian grass, shown here, and big bluestem together form the classic components of Eastern tallgrass prairie. Moisture-seeking roots may penetrate eight feet into the ground.

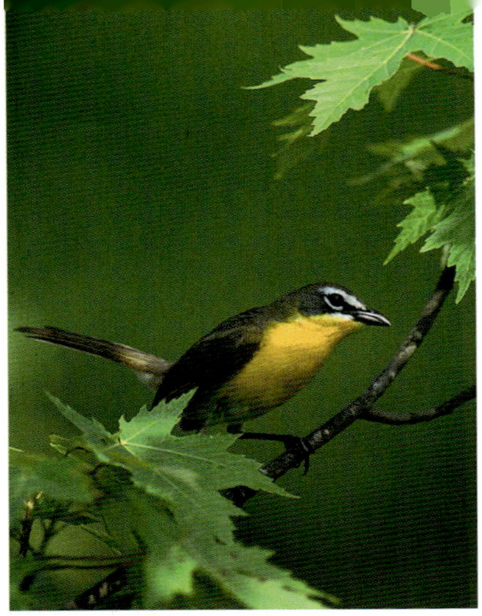

numerous American woodcock create a spectacle. At dusk, these long-billed sandpipers begin their odd courtship rituals. The male stakes out a patch of turf and emits short nasal *peent* calls. Eventually he launches skyward, circling high above the female observing from the ground, all the while emitting peculiar chirping sounds. For the finale, the woodcock drops rapidly earthward, creating squeaky chirping sounds from air rushing through specially shaped primary wing feathers.

In 1998 a new species of moth, *Spinopogon resthavenensis*, was discovered in prairie remnants at Resthaven Wildlife Area. It is drab and obscure—easily overlooked, and few lepidopterists would have recognized it as different. It tells a story, though; even in a well-explored place like Ohio, flora and fauna remain undiscovered and unstudied.

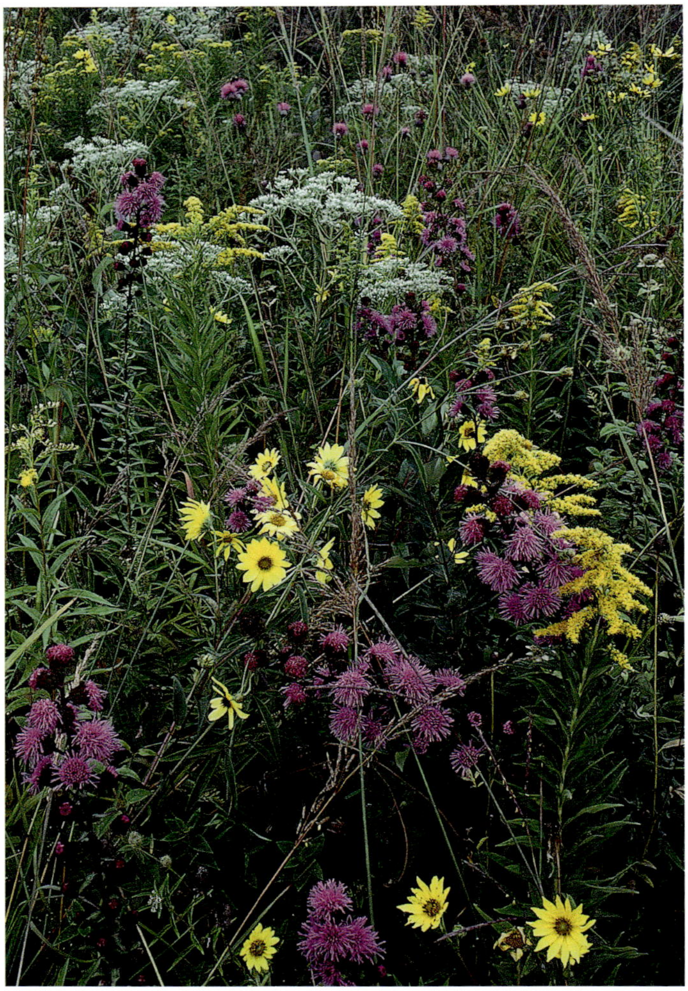

Top: Numbers of yellow-breasted chats, our largest warbler species, inhabit Resthaven Wildlife Area's prairie openings. These avian oddities have one of the largest repertoires of any songbird.

Left: Blazing-star, goldenrod, boneset, and sunflowers create a colorful palette amongst prairie grasses in late summer at Resthaven Wildlife Area.

Opposite: White lady's-slippers need fire to maintain a suitable habitat. The Ohio Division of Wildlife manages for this rare orchid by periodically burning the meadows in which they occur.

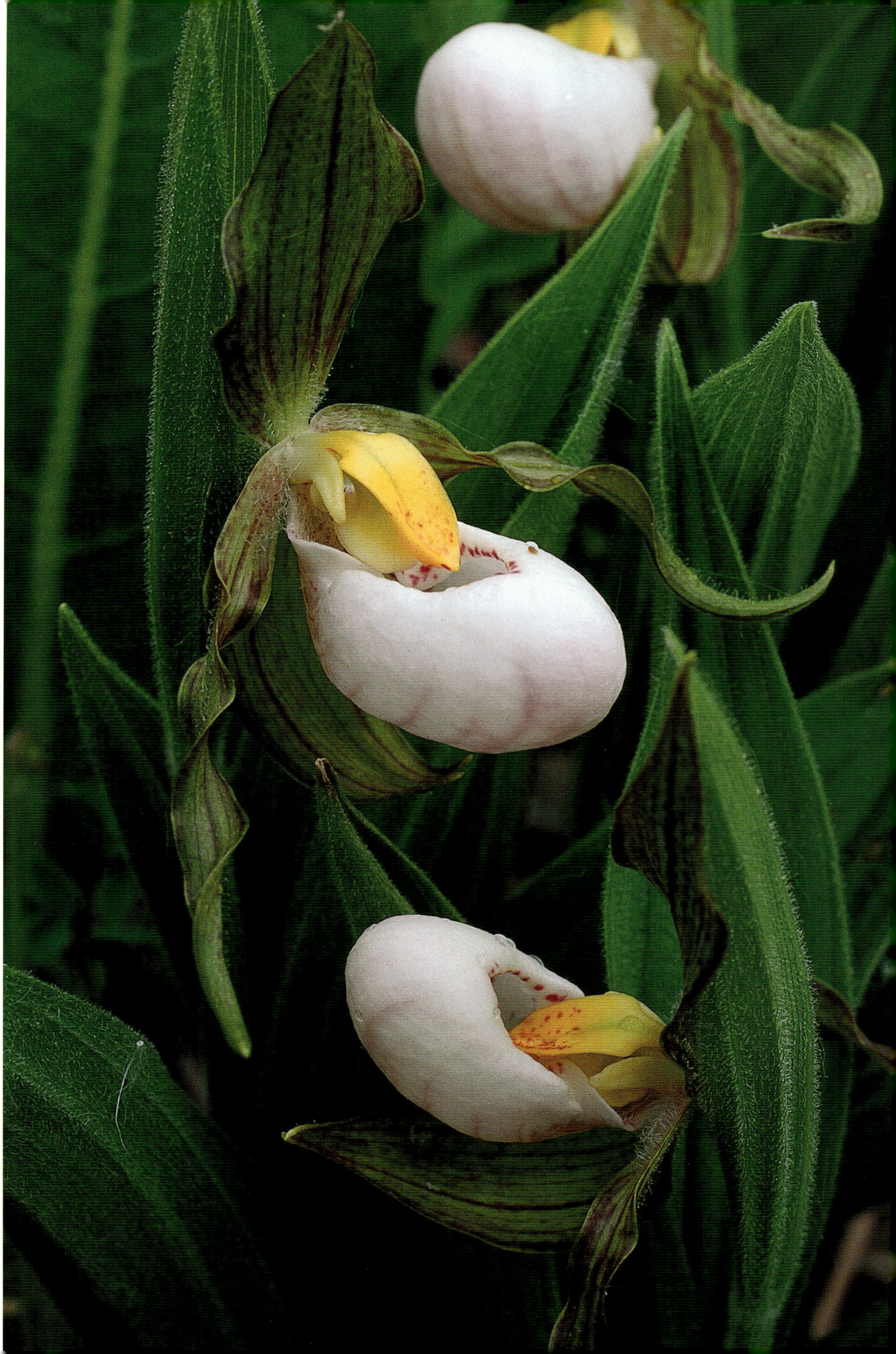

Glaciated Allegheny Plateaus

Buzzards Roost Nature Preserve (18) • Fern Lake and Lake Kelso (19) • Gott Fen State Nature Preserve (20) • Grand River Terraces (21) • Hell Hollow Wilderness Area/Paine Creek Watershed (22) • Johnson Woods State Nature Preserve (25) • Kent Bog State Nature Preserve (24) • Killbuck Marsh Wildlife Area (23) • Mantua Bog State Nature Preserve (26) • Morgan Swamp (27) • Singer Lake Bog (28) • Triangle Lake Bog State Nature Preserve (29)

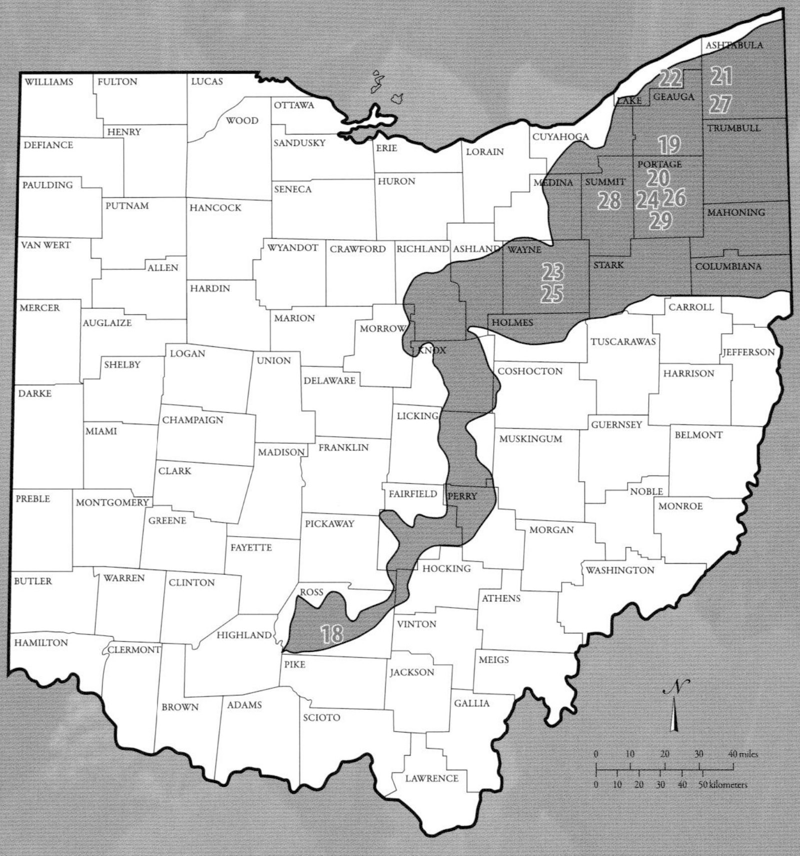

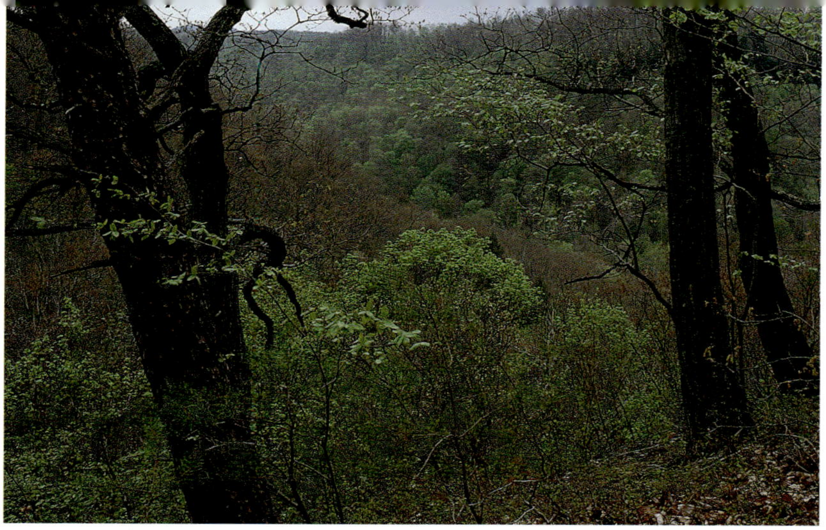

Buzzards Roost Nature Preserve

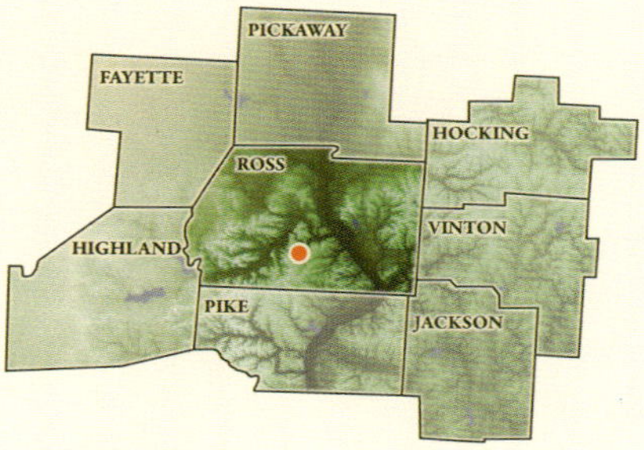

ACREAGE: Approximately 1,200

OWNERSHIP: Ross County Park District

ACCESS: Best entered at the western terminus of Red Bird Lane off Polk Hollow Road; take Polk Hollow Rd. south from U.S. Route 50 just west of Chillicothe.

WEB SITE RESOURCES: http://www.rosscountyparkdistrict.com/

http://www.svbnc.org/Earl%20H.%20Barnhart%20Nature%20Preserve.html

Top: Heavily forested bluffs and slopes of Paint Creek create a rugged wilderness setting just outside of Chillicothe.

The Paint Creek Valley is a place of geological convergence. At Chillicothe, hills of the Allegheny Plateau begin their rise, and flat Illinoian and Wisconsin till plains—remnants of our last glaciers—trickle to a stop just west of town. Perhaps most conspicuous are massive walls of black shale rising high above the valley. The largest of these are Alum Cliffs and Copperas Mountain, rising 250 and 550 feet, respectively, exposing massive banks of shale from several geologic epochs.

In the midst of this geological wonderland is the more than twelve-hundred-acre Buzzards Roost Nature Preserve. A place of diverse forests, the site—acquired and protected only in the past five years—is not well known to outsiders. As would be expected, such a mammoth, diverse landscape holds tremendous biodiversity, and much remains to be discovered. A June 16, 2004, "Botanical Blitz" conducted by two dozen botanists found over 350 species of plants—in one day! Many were new records for the preserve. Over six hundred species of plants have been found so far, and eventually seven hundred or more species may be documented.

Ridgetops high above the Paint Creek Valley are timbered in black and chestnut oaks, along with a smattering of scarlet oak, black-gum, and shagbark hickory. As one proceeds downslope, soils become richer and more moist, and tulip tree, red oak, and American beech begin to dominate. Lush stream terraces are covered with box-elder, green ash, silver maple, and Eastern cottonwood.

Along with big, diverse woodlands come lots of birds. Basic biological data is still being collected, and more additions will be added to Buzzards Roost's bounty, but the number of breeding birds currently exceeds eighty species. Many of these nesters are neotropical species that winter in the jungles of Central and South America. Included among their ranks are the perilously declining cerulean warbler, the stunning rose-breasted grosbeak, and a burst of understory brightness, the showy hooded warbler.

One of Ohio's best remaining streams is Paint Creek, and protection of buffering Buzzards Roost has surely helped protect its water quality. About forty-five species of fishes live in its waters, including pollution-sensitive rarities such as sand darter and the Northern madtom.The latter is a tiny catfish capable of delivering painful stings with its sharp pectoral spines. All told, nineteen species of fishes listed as endangered, threatened, of special interest, or under investigation have been found in Paint Creek.

Buzzards Roost is on the western edge of the unglaciated Allegheny Plateau. Trees tolerant of poor, acidic ridgetop soils like chestnut oak range no farther west, where limestone bedrock pushes close to the surface and sweetens the soil. This preserve is also one of our largest contiguous tracts of woodland in Ohio's central region. Visitors can experience a taste of the vast forests that cloaked presettlement Ohio and interesting flora and fauna that inhabit them.

One of six Ohio swallowtails, Eastern tiger swallowtails often course high in forest canopies, laying eggs and dipping earthward to nectar at favored flowers.

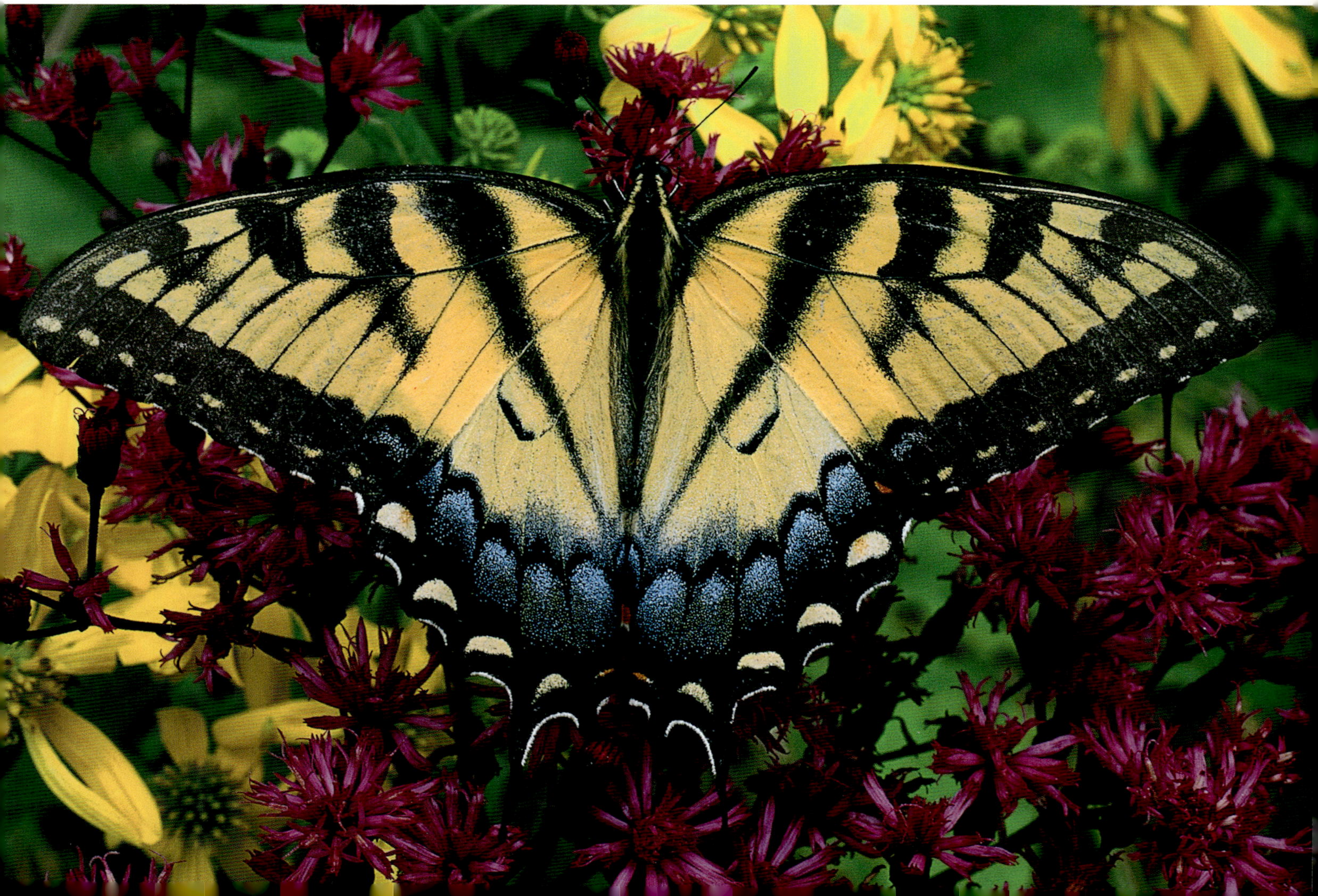

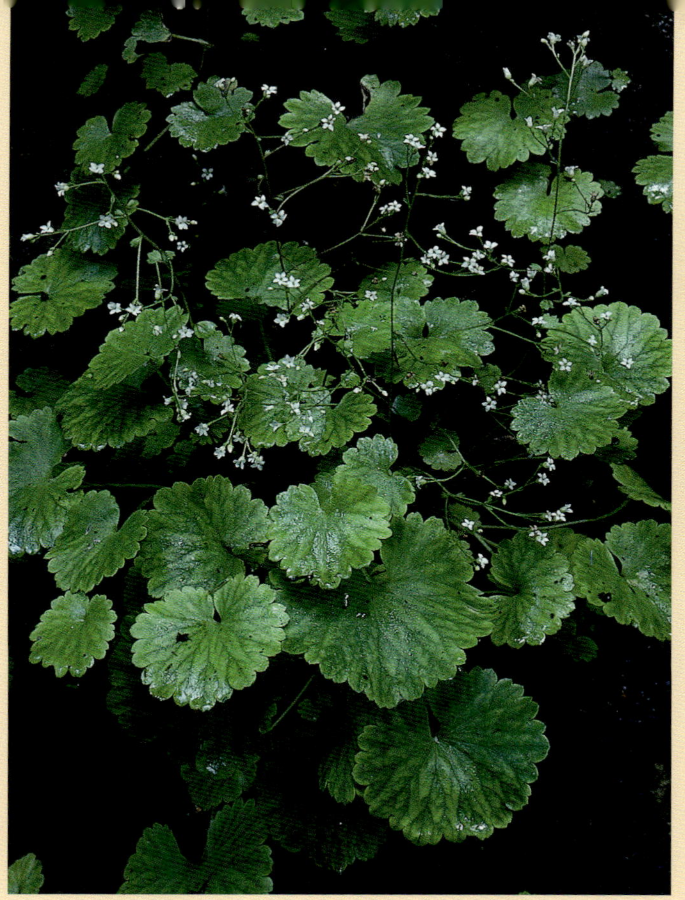

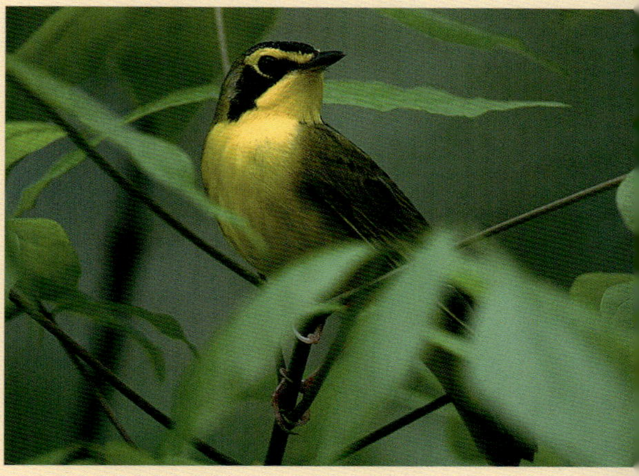

Furtive and hard to observe, male Kentucky warblers deliver a chanting song reminiscent of galloping horse hoof beats.

Top: The diminutive Sullivantia is saxicolous, or rock-dwelling. It occurs on shaded limestone cliffs along Paint Creek near Buzzards Roost. This member of the Saxifrage family has historical roots in the state. Famed botanist William Starling Sullivant of Columbus discovered Sullivantia in nearby Highland County, Ohio.

Right: Few birds are more secretive than the tiny, robin-sized Northern saw-whet owl. In 2003, a banding operation was launched at Buzzards Roost that has captured and banded over 110 owls to date.

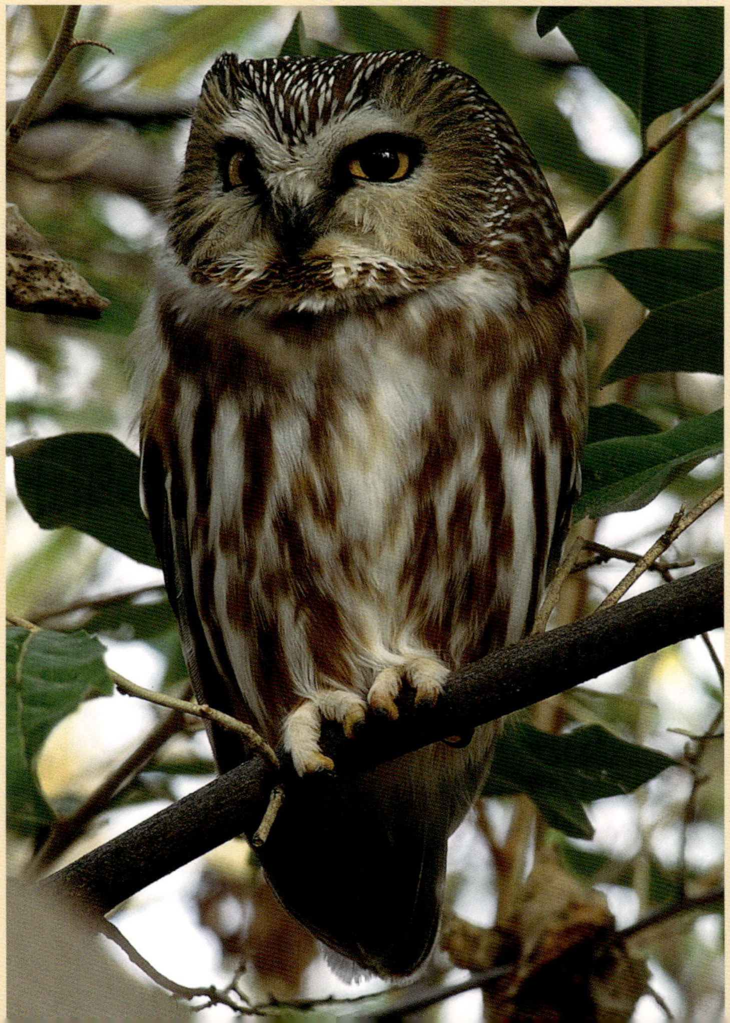

Fern Lake and Lake Kelso

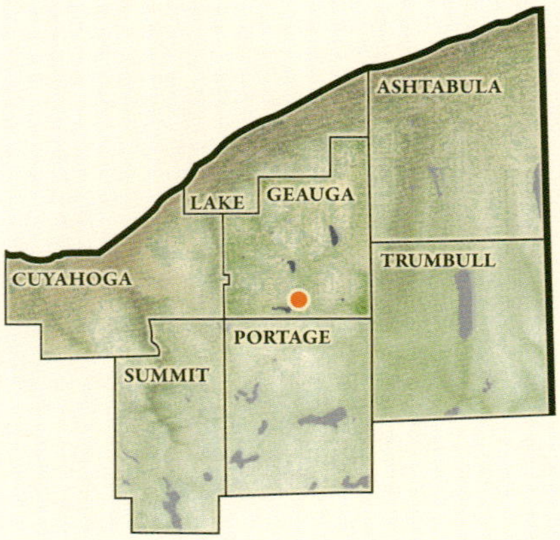

ACREAGE: 302

OWNERSHIP: Cleveland Museum of Natural History; Geauga Park District

ACCESS: Lake Kelso/Burton Wetlands open to public during posted hours; Fern Lake by permit through the Cleveland Museum of Natural History

WEB SITE RESOURCES:
http://www.cmnh.org/site/Conservation_NaturalAreas_Map_FernLakeBog.aspx

http://www.geaugaparkdistrict.org/parks/burtonwetlands.shtml

http://www.dnr.state.oh.us/dnap/location/burtonwetlands.htm

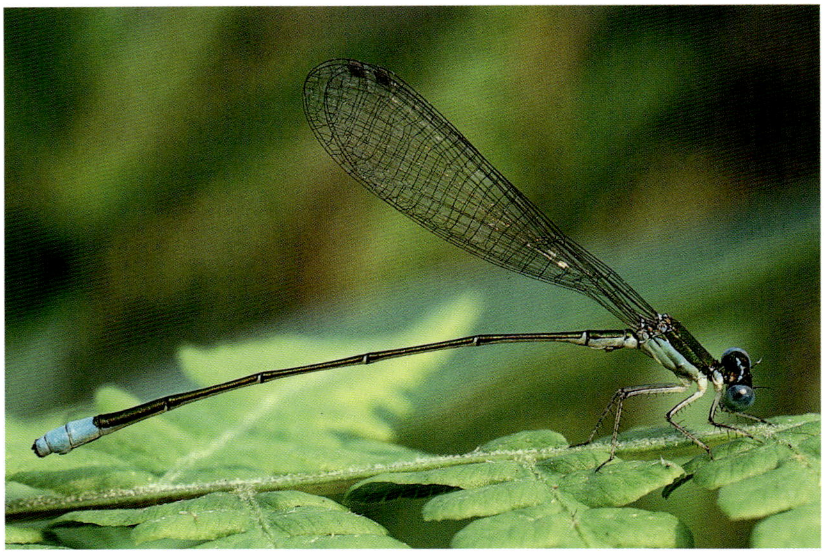

Pristine wetland complexes like Burton Wetlands/Lake Kelso and Fern Lake have become very rare in Ohio. Thanks to protection efforts by the Cleveland Museum of Natural History and Geauga County Park District, we can enjoy a northern bogscape in Ohio.

Glacial kettle bogs are one of our scarcest habitats. Prior to establishment of laws to protect wetlands, and the efforts of conservation organizations, most of our peatlands were either filled or mined for peat. A study published in 1992 showed that about 98 percent of Ohio's bogs and fens have been destroyed in the past century.

The 302 protected acres of Fern and Kelso lakes are part of a much larger system sometimes called the Cuyahoga Wetlands. Encompassing nearly one thousand acres, this region is dotted with small glacial kettles, swamp forests, and seepage meadows. Rarities abound, including the endangered American emerald dragonfly with its brilliant green eyes. This predatory insect is at the southern limit of its range here, with only a few known sites in Geauga and Portage counties.

At least twenty-nine species of state-listed rare flora have been documented at Fern and Kelso. Many are boreal relicts; travel to northern bog country and most would be common. About ten thousand years ago, in the wake of the retreating Wisconsin glacier, these plants would have been frequent in Ohio, too. The postglacier landscape was cool, boggy, and wide open—prime habitat for bog species. As the climate warmed, tundralike environments developed into deciduous forests. Today, specialized peatland flora only occurs in our best remaining bogs and fens.

Plants in the heath family dominate bog mats. A small shrub, leather-leaf, glosses up the carpets of *Sphagnum* moss in May, when it produces abundant racemes of small white flowers. Creeping along the surface is large cranberry, the same species elsewhere cultivated for its fruit. Of special note are small patches of the endangered small cranberry, only known from a few Ohio bogs. Of the twenty-one native Ohio heaths, seven are found only in peatlands and most are rare.

This is one of few Ohio sites that harbor pitcher-plants. Carnivorous plants like bladderworts and sundews, pitcher-plants are typical of bogs and evolved a meat-eating lifestyle to survive in nutrient-deficient substrates. Through different mechanisms, these plants lure insects to their modified leaves, which serve as killing traps. Once snared, the plant slowly digests the animal and extracts its proteins and nitrogen.

Opposite: A tiny carnivore, sphagnum sprites are less than an inch long. This damselfly is rare and local in Ohio peatlands.

Below: Beautiful Lake Kelso seems more apropos of northern Michigan or Canada than Ohio. Concentric bands of vegetation, layered in different plant communities based on substrate and water depth, ring the lake. Extending farthest into the depths is fragrant water-lily, which produces creamy-white floating blossoms.

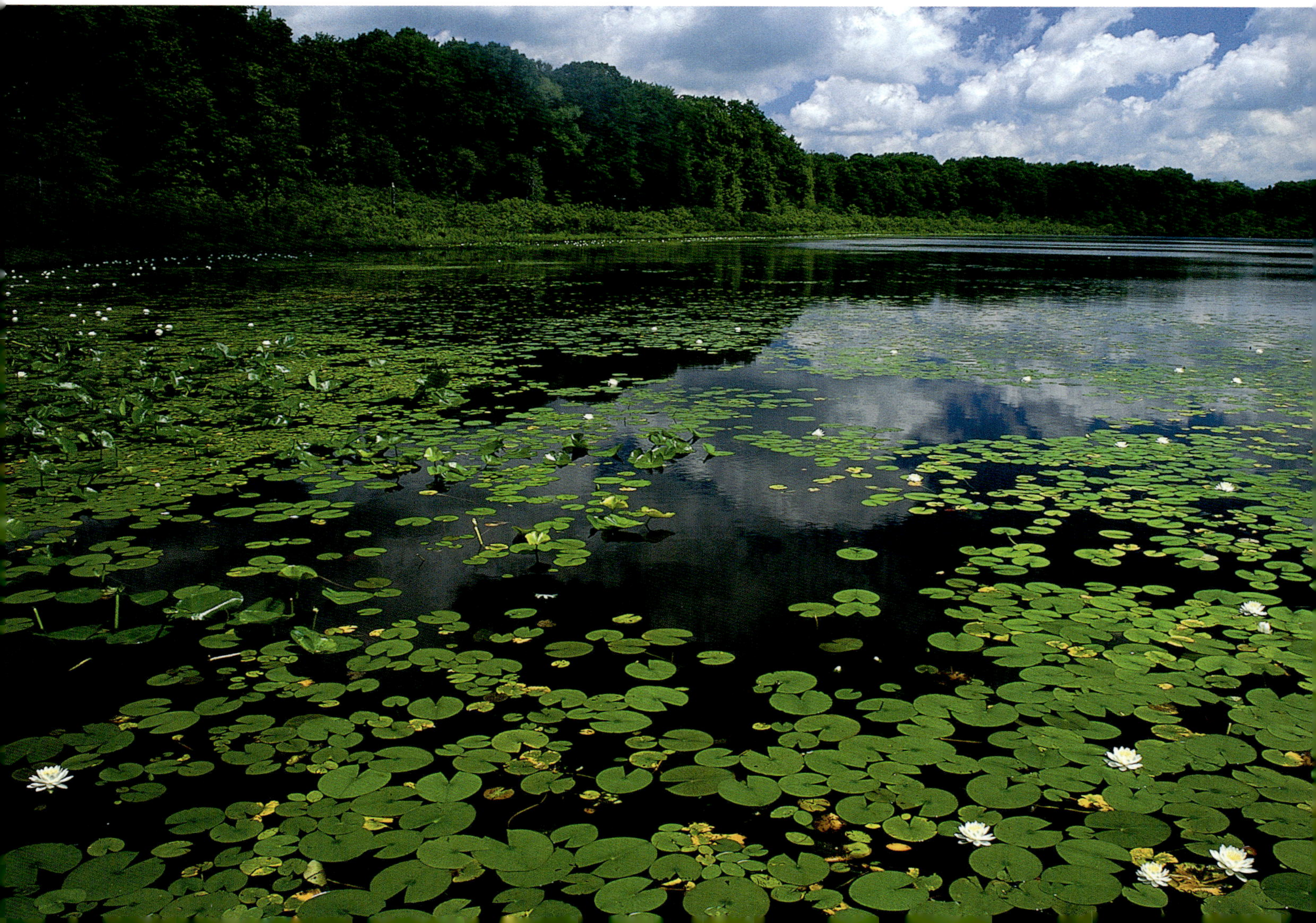

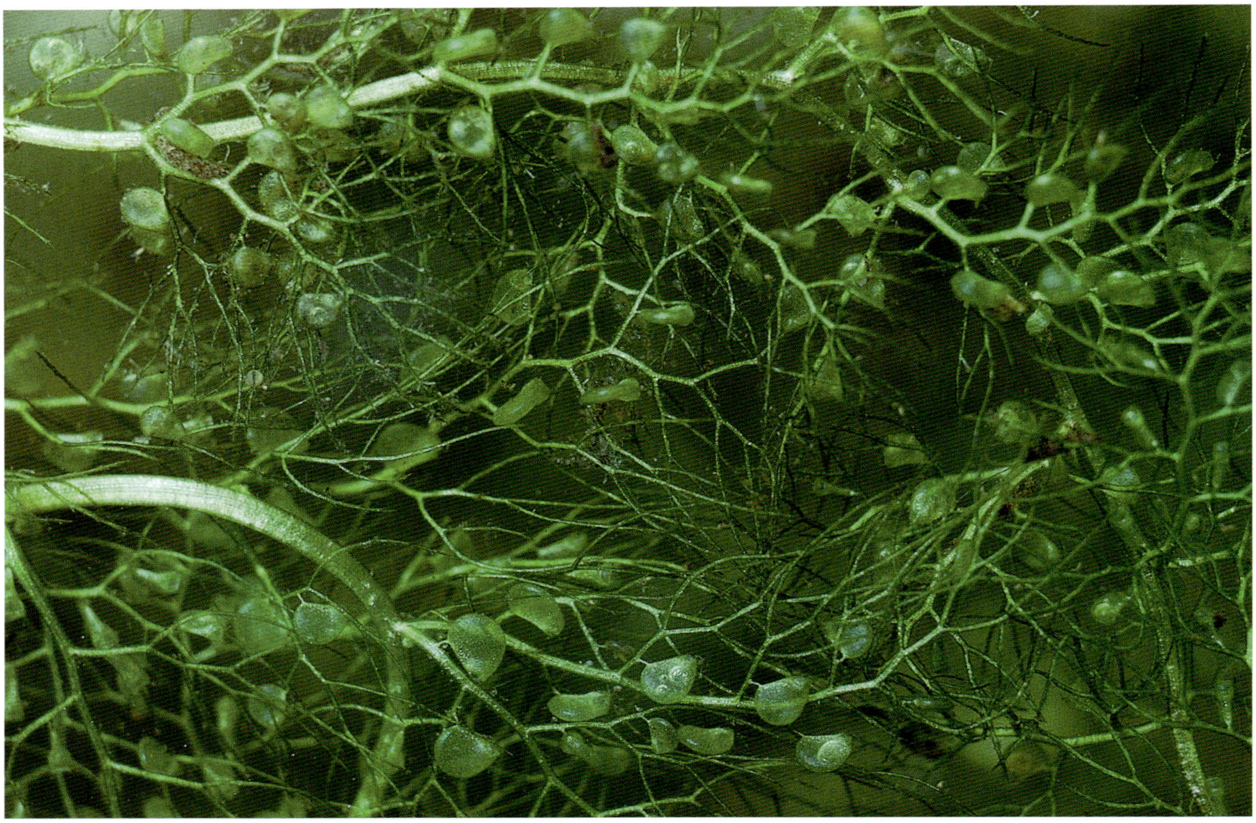

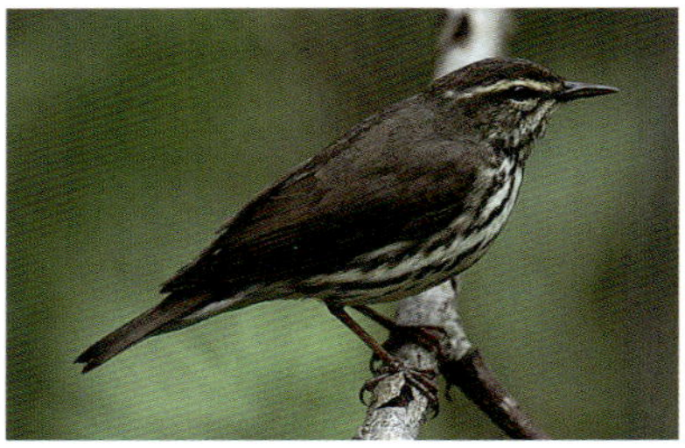

Top: Bladderworts are the largest group of carnivorous plants in the world. Six species occur in Ohio, including this common bladderwort. Their roots are heavily beset with tiny bladders, which trap aquatic animals.

Above: The waterthrush genus name *Seiurus* means to "wave tail." Northern waterthrushes incessantly bob their rear ends as they forage along the margins of boggy areas. They are common Ohio migrants, but very rare as breeders.

The City of Akron also owns land in the Cuyahoga Wetlands. As part of a visionary effort to protect a major water source, the Upper Cuyahoga River, this municipality began acquiring large chunks of land buffering the stream. Included in its holdings is the best white pine bog in Ohio. A soupy, hard-to-traverse morass, this rare wetland supports towering old-growth pines along with many rare plants.

Protection of Lakes Fern and Kelso and other areas of the Cuyahoga Wetlands has only been possible due to the efforts of several organizations. The Cleveland Museum of Natural History, The Nature Conservancy, Geauga County Park District, and the City of Akron have forged an alliance of partners working together to conserve one of Ohio's most biologically significant regions.

Fens are wetlands created by outflows of mineral-rich artesian springs. Water that has percolated through subterranean limestone aquifers, becoming infused with calcium bicarbonates, eventually is forced to the surface by hydrostatic pressure. Exploding from pores in the earth's surface, the chilled water rushes along small rivulets and saturates soils in the vicinity.

The end result is the fen meadow: lush, densely vegetated landscapes filled with specialized plants. The combination of cold root-zone temperatures, thoroughly saturated and highly alkaline soils, and sizzling aboveground summertime heat precludes most plants that didn't coevolve with fen habitats.

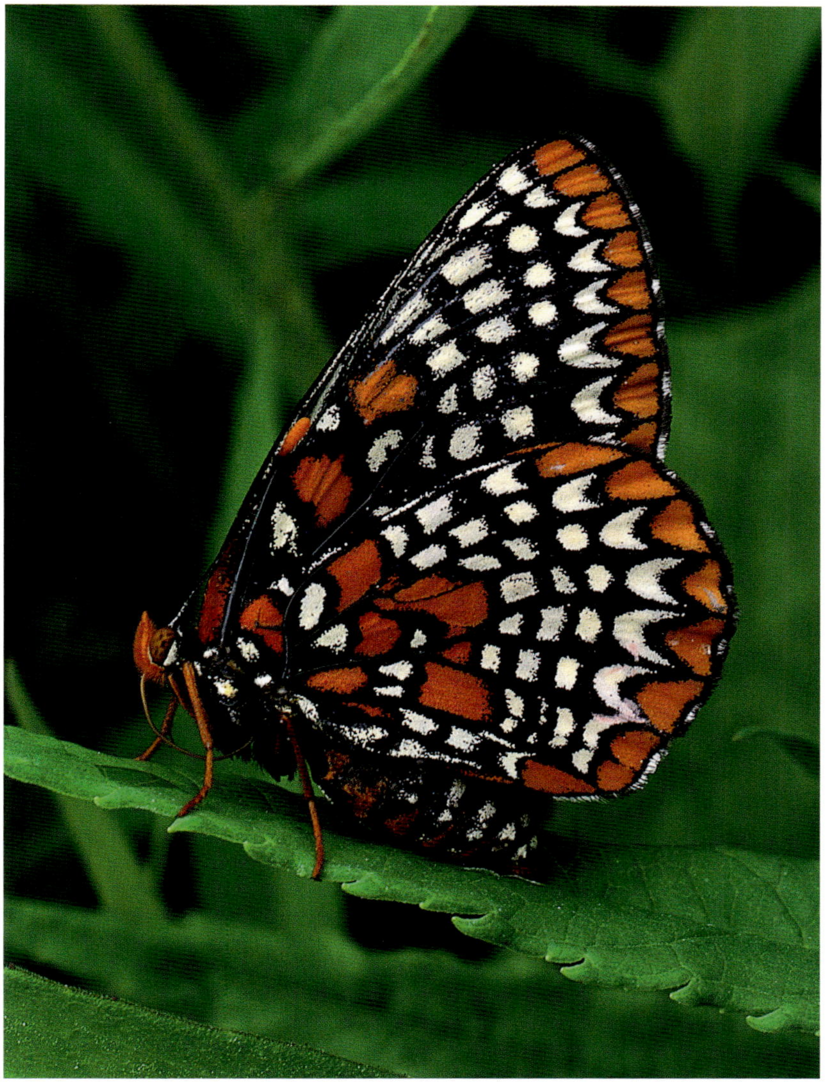

Gott Fen State Nature Preserve

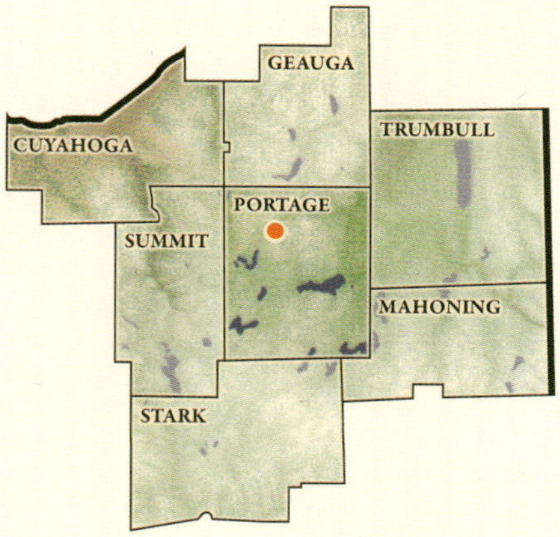

ACREAGE: 44

OWNERSHIP: Ohio Division of Natural Areas and Preserves

ACCESS: By permit only

WEB SITE RESOURCES: http://www.dnr.state.oh.us/dnap/permitonly.htm

Baltimore checkerspots can be found in fens and wet meadows. The young caterpillars band together within a communally constructed silken, bagwormlike nest.

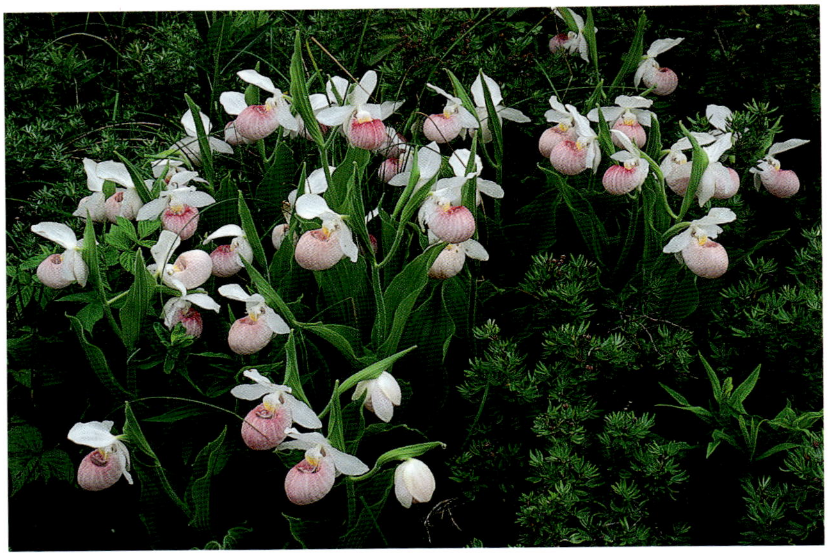

Above: Showy lady's-slippers are North America's largest terrestrial orchid. The huge, saclike flower has a large opening that permits bumblebees to enter and pollinate. Look, but don't touch; handling the foliage can cause dermatitis similar to poison ivy. This orchid is threatened in Ohio; only five populations are known.

Opposite: The diminutive grass-of-parnassus brightens the fen meadows in mid to late summer. This saxifrage species is habitat-specific. It only occurs in saturated alkaline soils.

Some ecologists classify Ohio fens into two types: prairie fens and boreal fens. Gott Fen is among the latter; many of its plants are northerners at the southern limits of their ranges in Ohio.

To a botanist, few Ohio habitats are more interesting than fens. These wetlands are loaded with rarities, many of which grow in no other habitat. Gott Fen has at least twenty-three species of state-listed rare plants. Perhaps the most significant is bayberry, a small shrub with whitish waxy berries. Bayberry is endangered and only occurs in one other nearby locale. Bayberries, including this species, are sometimes used in candle-making.

Another noteworthy rarity is our smallest dogwood, bunchberry. This tiny shrublet stands but six inches tall and is endangered in Ohio. While wide-ranging and abundant across boreal forests of Canada and the northern United States, it barely reaches our latitude. It is thought that bunchberry is intolerant of soil temperatures with an average mean of greater than sixty-five degrees; thus most Ohio habitats are too warm.

Adequately protecting fens is difficult. While the fen proper can be purchased and managed, that is but part of the equation. Groundwater that feeds the hydrology of these rare wetlands is derived from expansive underground aquifers, and waters that eventually percolate into the fen may have entered the ground miles away. Thus, alteration of far-off habitat may adversely impact the quality and quantity of water feeding the fen.

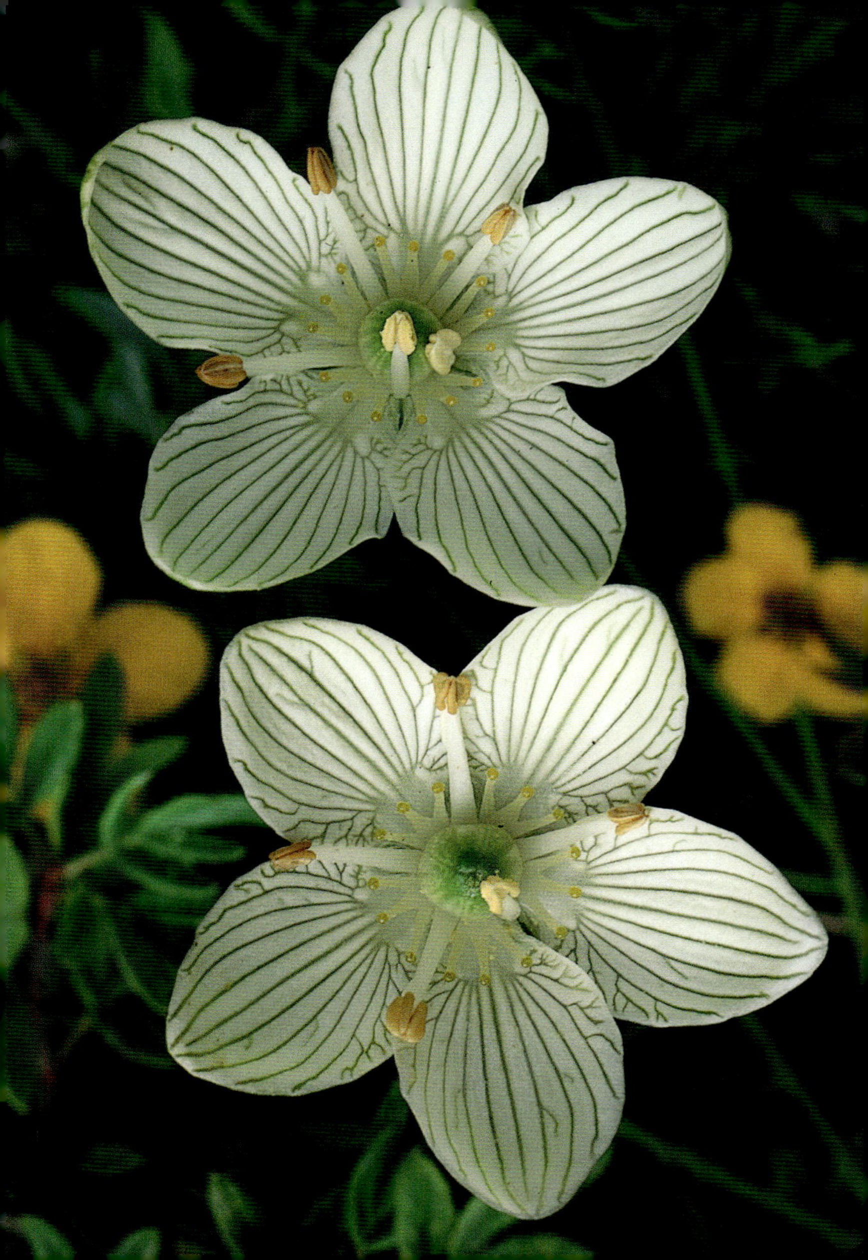

Grand River Terraces

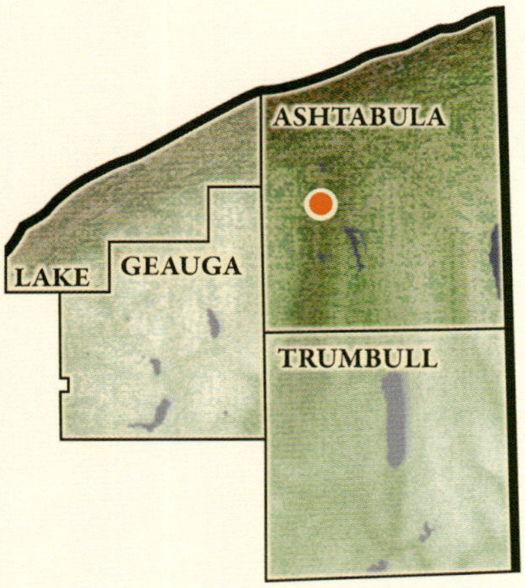

ACREAGE: 745

OWNERSHIP: Cleveland Museum of Natural History

ACCESS: By permit only

WEB SITE RESOURCES: http://www.cmnh.org/site/Conservation_NaturalAreas_Map_GrandRiverTerraces.aspx

Opposite top: Turk's-cap lilies are amongst Ohio's most spectacular wildflowers. Reminiscent of a Turkish turban, this lily occurs in alluvial soils of floodplain forests.

Opposite bottom: Grand River Terraces is an excellent example of a Lake Plain hemlock swamp. Lush beds of cinnamon and royal ferns grow on hummocks surrounding the woodland pools.

During the Pleistocene, misshapen lobes of ice jagged off the advancing face of the Wisconsin glacier, drilling valleys into the rugged landscape of northeast Ohio. When these icy spurs shoved against impervious sandstones not far south of Lake Erie, glacial progress halted. The results were valleys eroded into softer shales abutted by more resistant rocks. The odd, sinuous course of the Grand River is the result of this glaciation; the river was forced into a valley created by glacial scouring.

Located in our largest county, Ashtabula, the Grand River Terraces stretches for about a mile along the Grand River. Acquisition began in 1982 when the Cleveland Museum of Natural History took possession of the original seven hundred acres. The property contains excellent examples of floodplain forests and vernal pools. The finest hemlock swamp in Ohio occurs here.

Hemlocks are northern conifers and, in Ohio, almost always are found on steep slopes of cool ravines. Their occurrence in flat, swampy ground is rare, and as befits such an unusual habitat, a number of rare plants also are found in hemlock swamps. Among the scarcest is an odd member of the rose family, robin-run-away. This unroselike oddity creeps low over the ground, and the terraces host Ohio's largest population.

Five species of orchids have been recorded thus far. One of the most interesting is crane-fly orchid. The insect-shaped blooms are noctodorous—they produce a fragrance at night. The aroma attracts certain moths, and the shape of the flower forces pollen into contact with their fuzzy eyes.

Woodland pools harbor noteworthy breeding populations of at least three species of mole salamanders. Jefferson, silvery, and spotted salamanders migrate to these vernal wetlands in early spring, mate, drop egg masses, and return to subterranean haunts where they spend most of the year.

The streamside habitats of the Grand River Terraces directly protect the water quality of one of Ohio's most pristine rivers. The ninety-eight-mile-long Grand River drains 712 square miles of northeast Ohio, and a thirty-three-mile section has been named a state scenic river. Twenty-three miles are designated as wild, one of few Ohio streams to merit that classification.

Because corridors along the Grand River have been protected, many sensitive aquatic organisms still flourish. Riffles in the terraces support thriving freshwater mussel beds that include at least eleven

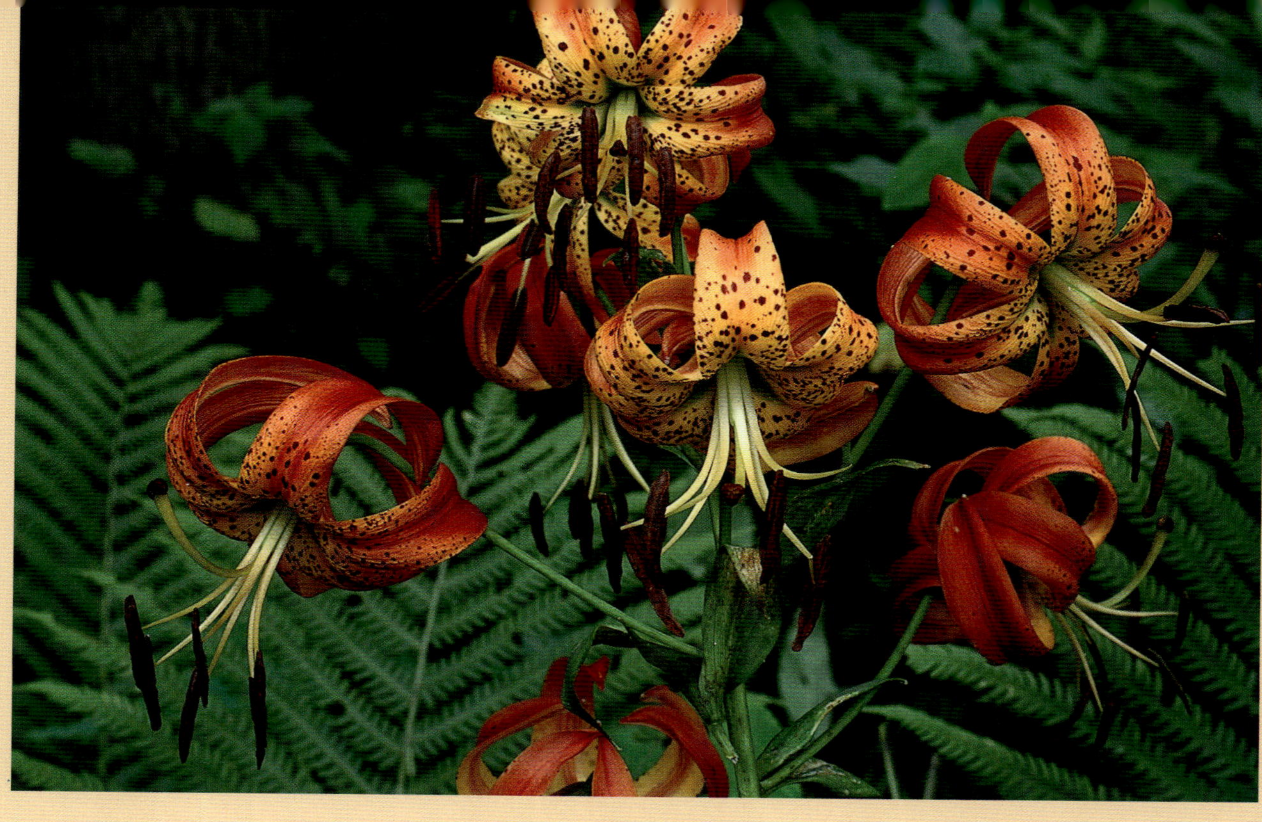

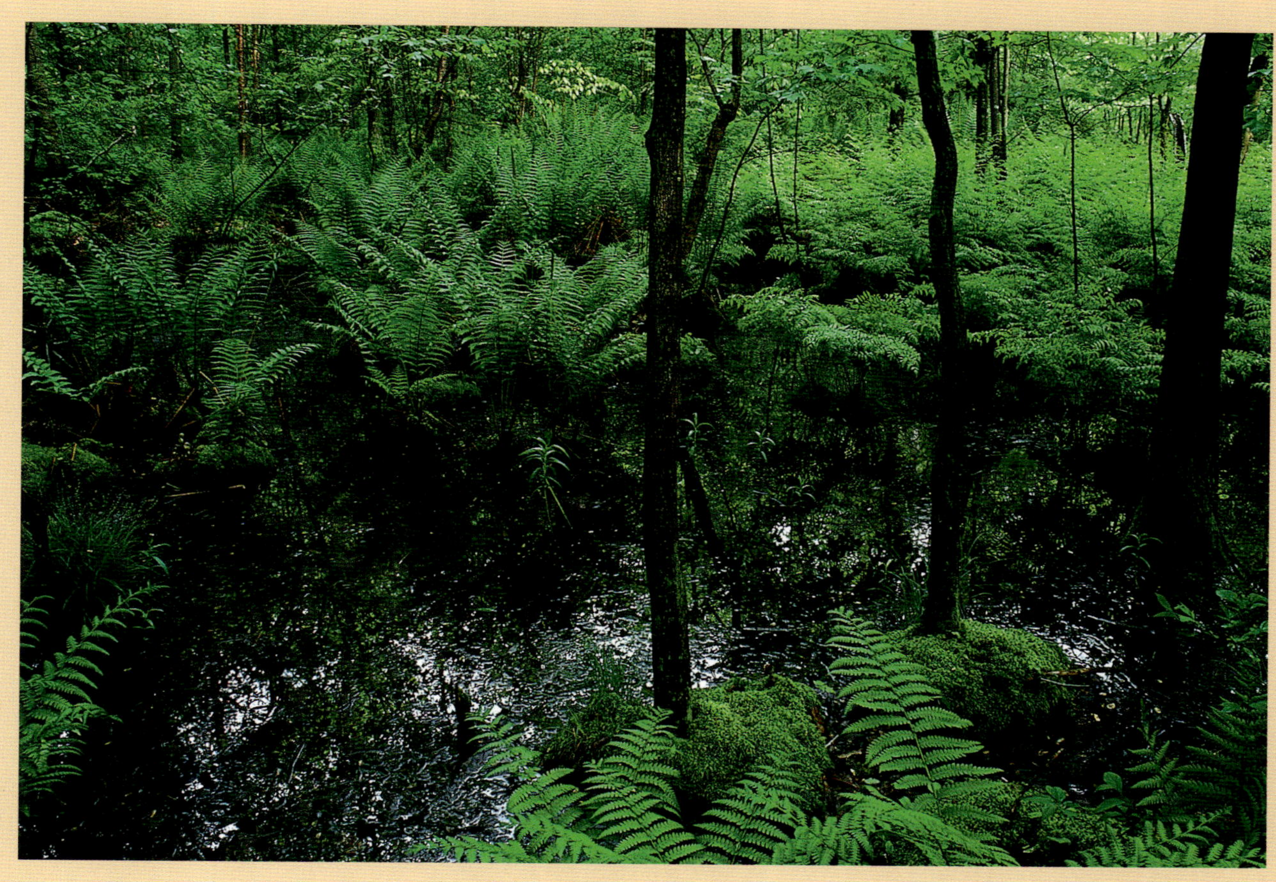

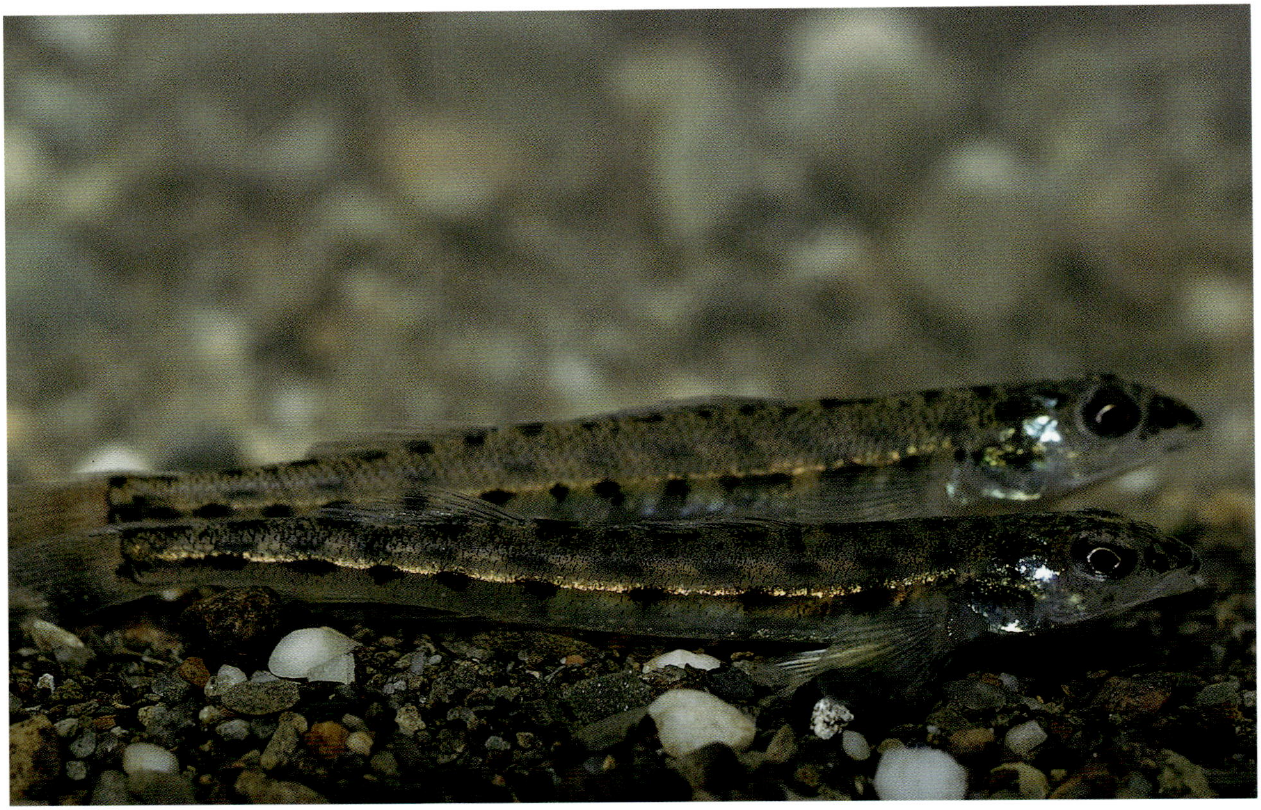

Sand darters are found on clean, sandy stream bottoms. This darter has declined significantly in Ohio and is listed as a species of concern. Excessive siltation of our waterways has eliminated this species from many of our streams.

species, among them the rare black sandshell and round pigtoe. Not far downstream of the terraces, a bizarre aquatic plant, riverweed, grows in midriver rapids. This mosslike plant is extremely sensitive to water-quality degradation, and the Grand River supports Ohio's only population.

In total, the Grand River Terraces and other riverine habitats in the watershed support at least fifty state-listed rare plants, about one hundred species of breeding birds, twenty-eight species of amphibians and reptiles, and seventy-four fish species—incredible biodiversity by any standard, making the Grand River Terraces and the Grand River one of Ohio's most significant natural resources.

Steep, deep ravines in the Paine Creek Watershed harbor boreal flora unlike most of Ohio. Lying in the middle of the Snow Belt, it receives an average of 106 inches of snow annually. Habitats here resemble coniferous woodlands of Canada. Gorges such as Hell Hollow host dense stands of hemlock, and understory shrubs include rare northern species such as hobblebush. Adventurers seeking the cool ravine bottoms can hike down a 262-step wooden staircase.

Some of the standout rarities of the region are birds. A potpourri of unusual breeders occupies hemlock glens in Hell Hollow, such as black-throated green; Canada, magnolia, and Blackburnian warblers; hermit thrush; and winter wren. The last, a four-inch-long ball of feathers, sings one of the most complex songs in the bird world. This tune seems to perfectly match the shadowy beauty of the hemlock gorges.

Because streamside habitats have been well protected in the Paine Creek drainage, aquatic habitats are outstanding. One of two species of rare dragonflies found along local watercourses is Uhler's sundragon, an Ohio endangered species. Rather dull except for brilliant green eyes

Hell Hollow Wilderness Area/ Paine Creek Watershed

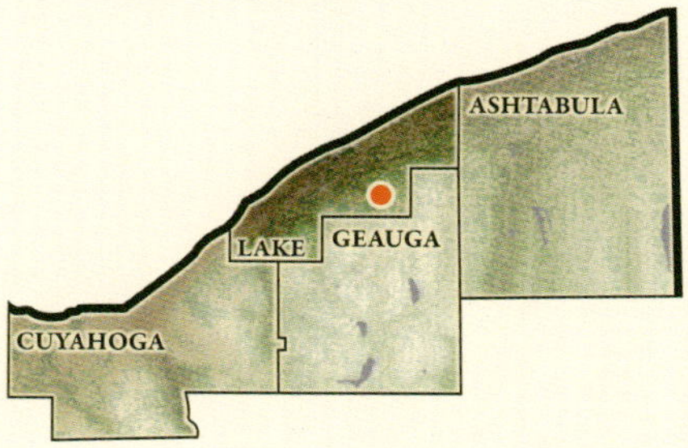

ACREAGE: 783

OWNERSHIP: Lake County Metroparks

ACCESS: Open to public during posted hours

WEB SITE RESOURCES: http://lakemetroparks.com/select-park/hell-hollow.shtml

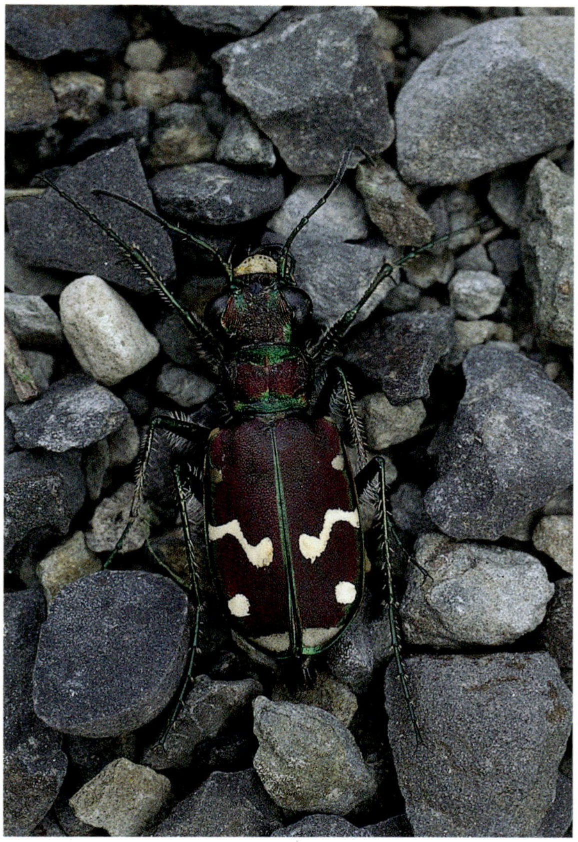

The claybank tiger beetle is one of nineteen species of tiger beetle occurring in Ohio. Shale banks along Paine Creek support the only known Ohio populations of this rare insect.

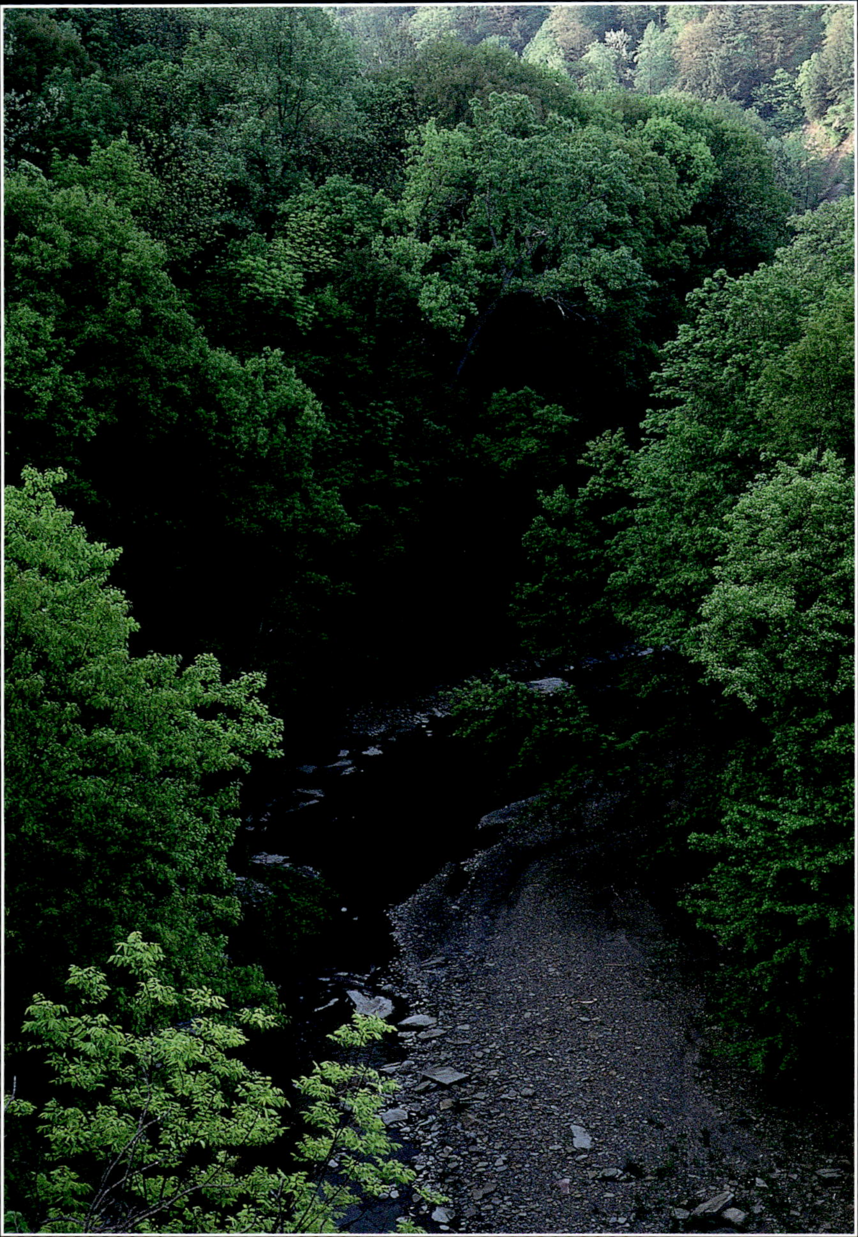

like others in the emerald family, adults often shun streamside haunts and forage in sunny forest openings.

Riffle snaketails, so named for their serpentlike abdomen, are intimately tied to streams. The adults loaf on rocks in the current, making regular patrols over the water to capture small insects and ward off rival males. This dragonfly is threatened in Ohio, and Paine Creek supports one of few populations.

Dragonflies serve as excellent barometers of stream health. Although adults aren't necessarily water-dependent, larvae are. Often called nymphs, dragonfly larvae spend anywhere from a few months to several years in the water, feeding and growing. Like miniscule aliens from a horror movie, they sport ejectable mandibles. In a flash, the nymph lunges its jaw forward an extraordinary distance and seizes unsuspecting prey.

Some 360 million years ago, conditions here were far more aquatic than today, although water remains a conspicuous feature. During the Devonian period, a shallow sea covered the land, and as it receded, shale deposits were left behind. Today, an impressive bank of Chagrin Shale can be seen from Hell Hollow Wilderness Area. The Paine Creek Watershed also contains a number of fine waterfalls. About 250 million years ago the first dragonflies took wing, and members of the *Odonata* have probably been hunting the waters of Paine Creek's watershed ever since.

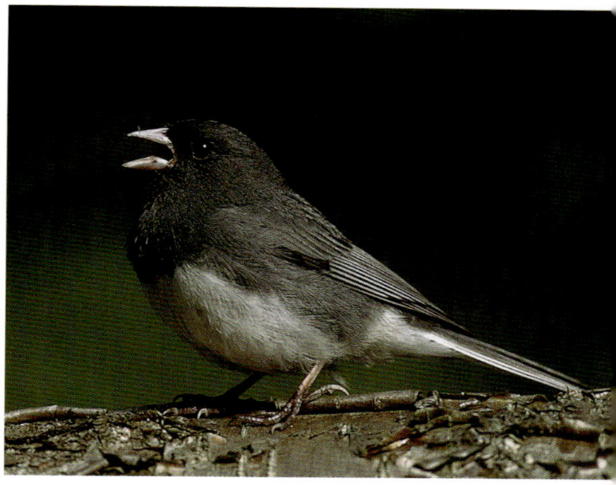

Opposite: Dense forests buffer the steep slopes along Paine Creek, protecting the water quality of the stream. Biodiversity is exceptionally high in this watershed because much of the adjacent lands have been protected.

Above: Breeding dark-eyed juncos are rare and locally distributed in northeastern Ohio. Hell Hollow's ravine supports many pairs of this species.

Bottom: Five species of dace inhabit Ohio streams. Breeding in clear pools, redside dace populations are found along the unglaciated Allegheny Plateau.

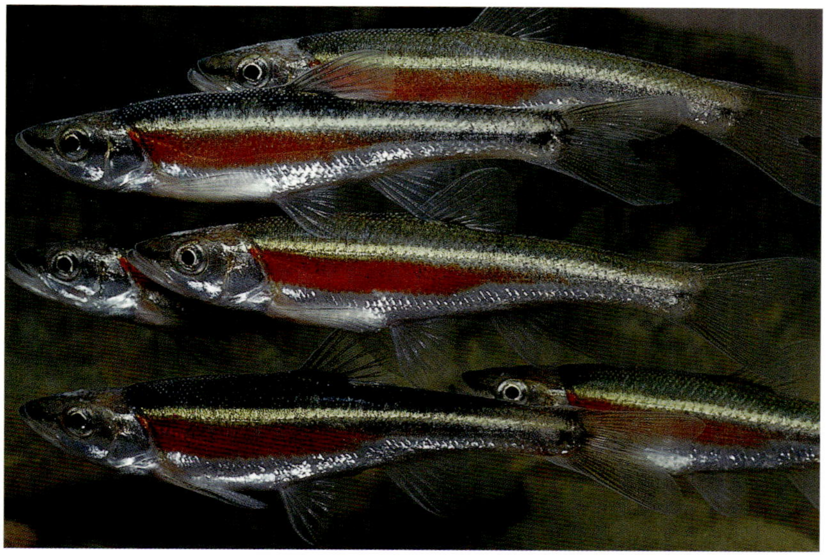

Johnson Woods State Nature Preserve

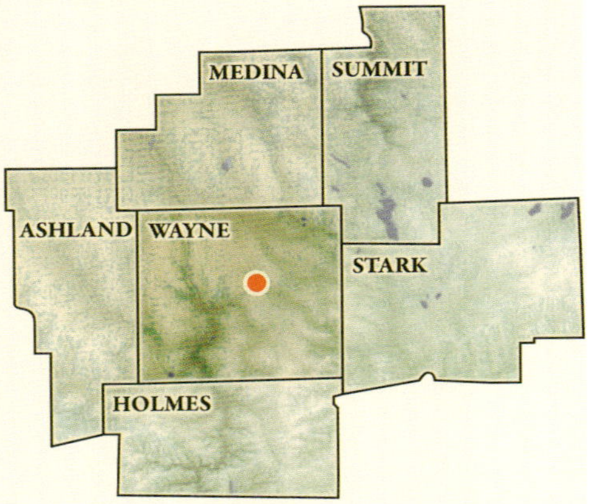

ACREAGE: 206

OWNERSHIP: Ohio Division of Natural Areas and Preserves

ACCESS: Open to the public during posted hours

WEB SITE RESOURCES: http://www.dnr.state.oh.us/dnap/location/johnson_woods/tabid/898/Default.aspx

Opposite: We owe the existence of Johnson Woods to the generosity of Clela and Andrew Johnson and their family, who ensured that this 206-acre patch of primeval timber remained uncut. Many of the trees here are over 300 years old and tower to 120 feet in height in this outstanding woodlot.

There are precious few Ohio woods that harbor trees over three hundred years old, but this site is one. A walk through the giants at Johnson Woods is sure to inspire awe among visitors. Some of the trees climb well over 130 feet skyward and have girths the size of big dinner tables. Once, over 90 percent of Ohio was covered with similar forests.

The trees at Johnson Woods, old and mature when Ohio achieved statehood in 1803, stood firmly rooted when the first pioneers swept into the region. It's likely that gray wolves and elk—creatures long gone—once roamed in their shade. To Jacob Conrad, the original owner of Johnson Woods, such a landscape would have been mundane in 1823 when he made his claim. Little did he know that, less than two hundred years later, only a tiny fraction of a percent of Ohio's primeval timber would remain.

The largest trees are red and white oaks, and shagbark hickory. Some of these leviathans are near the end of their lifespan. As upland oaks and hickories gradually fade from the scene, faster-growing and shorter-lived species that flourished in the shady understory dominate. Sugar maple and American beech will one day be the principal tree species at Johnson Woods. This shift in plant communities is known as ecological succession and is inevitable. Fire was far more frequent in presettlement landscapes than is widely recognized. Burns did not affect thick-barked, heat-resistant oaks and killed softwoods like maple. People have a natural aversion to wildfires, though, and fire suppression has become the norm in modern times, accelerating the shift from oaks to maples.

Low-lying woodland pools are ringed with pin and swamp white oak, red maple, and green ash. In the wettest areas is an interesting shrub, buttonbush. A member of the coffee family, buttonbush is thick with balls of white flowers in summer. Butterflies, especially silver-spotted skippers, swarm to the blooms seeking nectar.

Visit Johnson Woods in early May, before leaves have yet to mask the true dimensions of the grand trees. The forest floor will be a carpet of wildflowers: yellow trout-lilies, large-flowered trillium, spring-beauties, wild geraniums, and dozens of others. Newly arrived neotropical birds like scarlet tanager, yellow-throated vireo, and great crested flycatchers send down their melodies from the canopy. Then imagine that thirty-seven thousand square miles of Ohio looked like this only two hundred years ago.

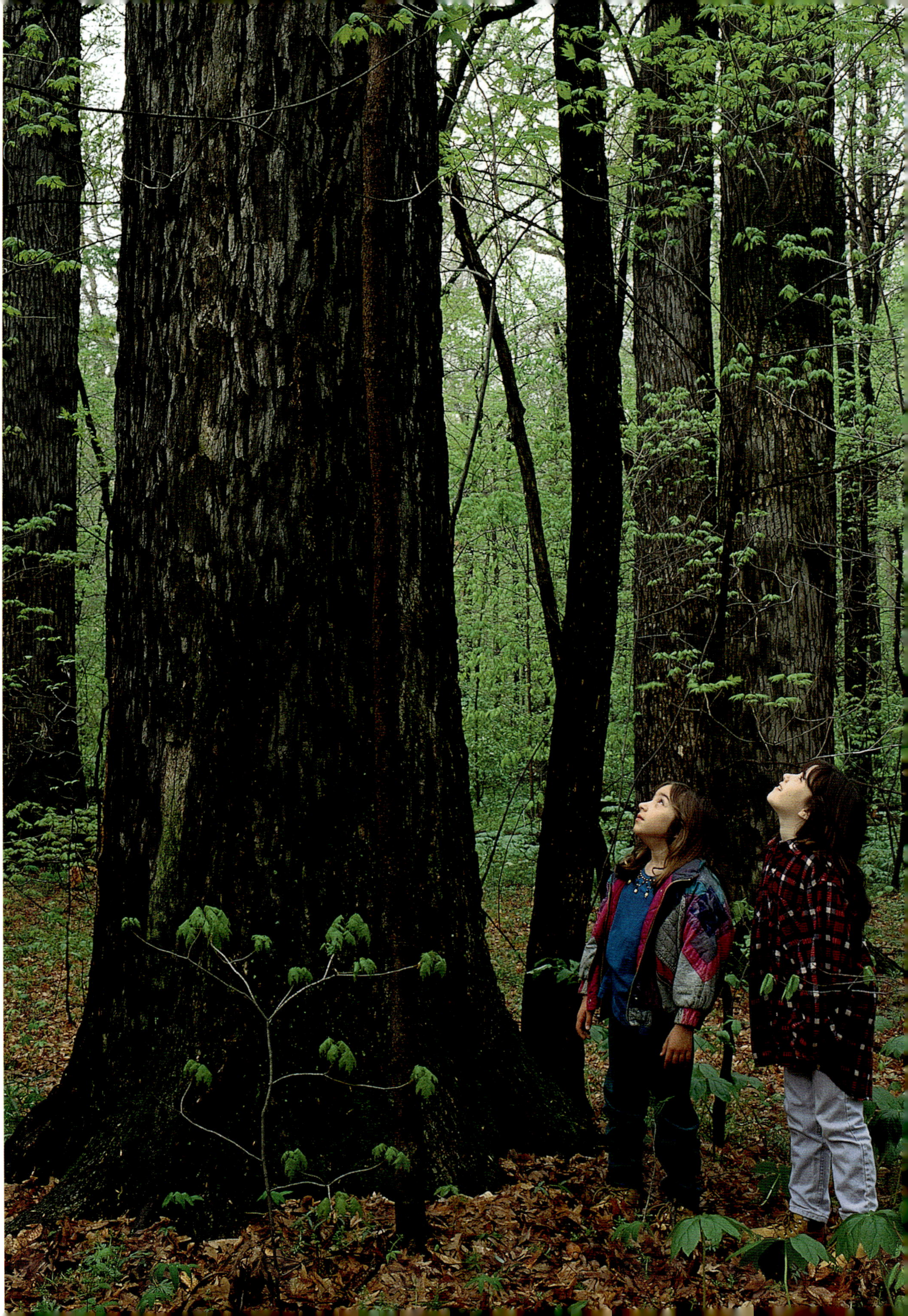

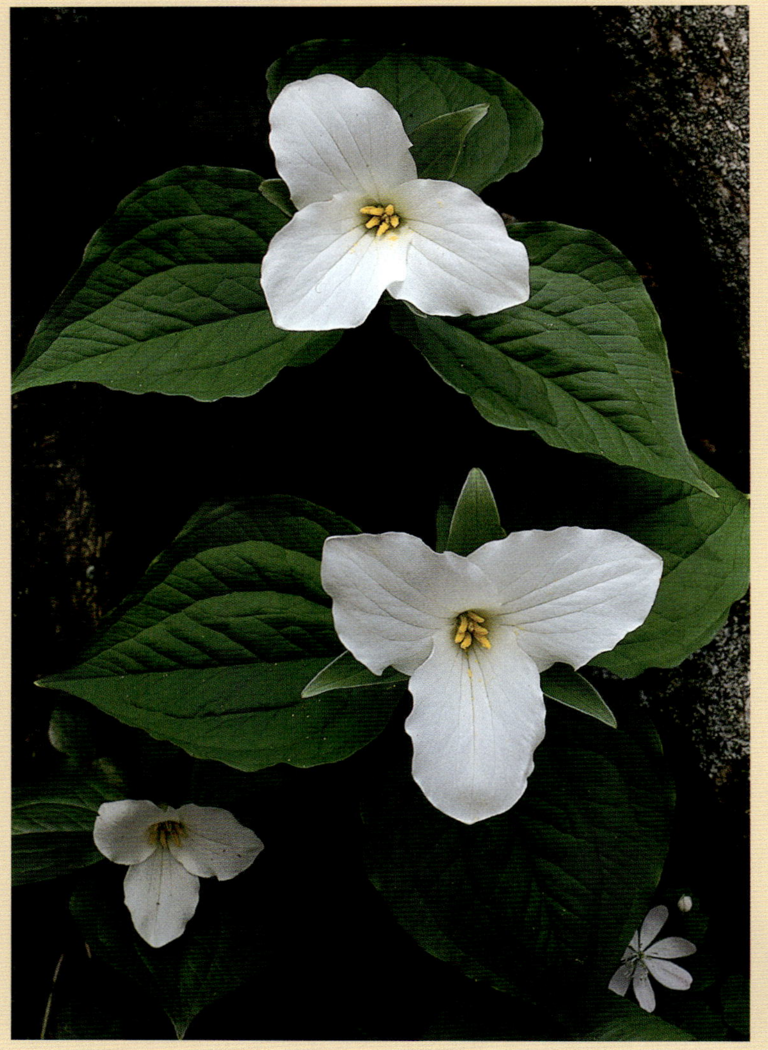

Left: Large-flowered trillium was declared the Ohio state wildflower in 1986. Fresh blooms are creamy white; as the flower ages, it becomes tinged with pink. White-tailed deer enjoy snacking on trillium and other lilies and have caused localized population declines.

Below: This is the largest woodpecker remaining in North America. It favors mature forests. Pileated woodpeckers excavate large, oval-shaped nest holes that are utilized by many other animals.

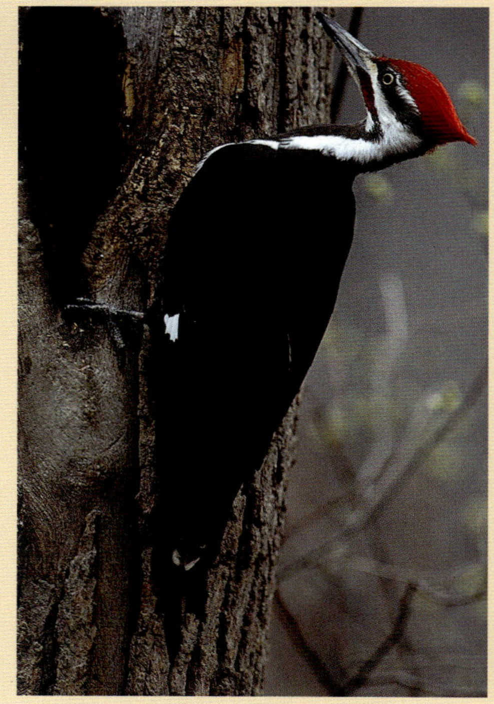

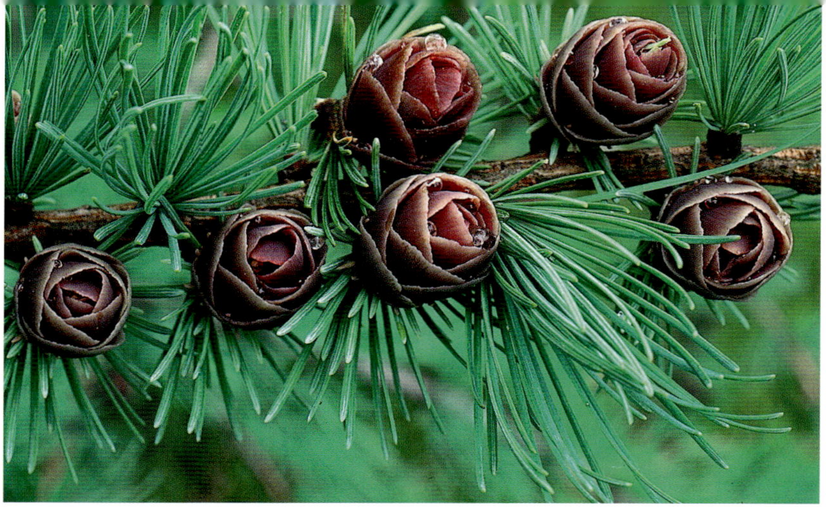

Top: Before dropping from the trees in late fall, the needles of tamarack, our only deciduous conifer, turn a striking golden yellow. In Ohio, tamarack is confined to peatlands. The small cones shown here are valuable food for squirrels and birds.

Kent Bog State Nature Preserve

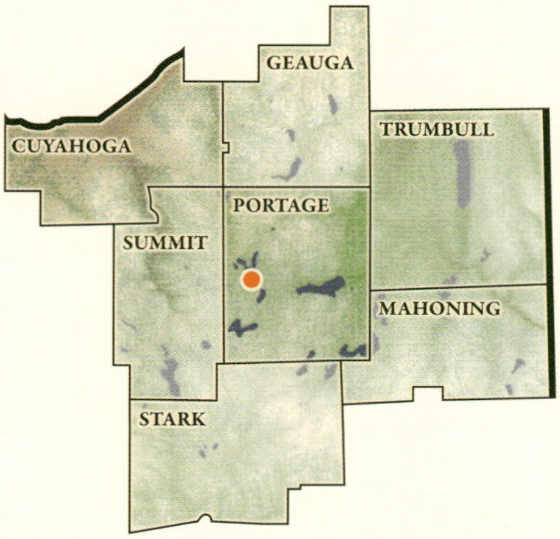

ACREAGE: 42

OWNERSHIP: Ohio Division of Natural Areas and Preserves

ACCESS: Open to the public during posted hours

WEB SITE RESOURCES: http://www.dnr.state.oh.us/dnap/location/KentBog/tabid/900/Default.aspx

Ohio bogs were created about twelve thousand years ago, spawned during the retreat of the Wisconsin glacier. Enormous ice chunks calved from the glacier's face created gigantic divots in soft, mucky soil, and these depressions filled with water.

Over time, plants colonized the kettle lakes, and *Sphagnum* mosses formed thick layers. Eventually, *Sphagnum* proliferate to the point where it became the dominant substrate. As a byproduct of its growth, these spongy mosses exude acidity, thus influencing the types of plants that can grow within the bog. Most plants aren't acid-tolerant, and many of those that are tend to be rare in Ohio and confined to bogs.

After thousands of years, plants eventually squeeze out open water and overtake the kettle lake. The amount of time required for this ecological process varies and is dependent upon the size and depth of the lake, but ultimately plants will win out.

Kent Bog is far along in its life. No open water remains and the bog mat is thickly overgrown with tamarack and gray birch. The gnarly tamarack forest at Kent Bog is the largest stand found this far south. Interesting shrubs form dense thickets in the bog, including catberry and winterberry, two of the three native Ohio hollies. Eight-foot-tall hedges of highbush blueberry produce bounties of sweet berries in summer. The most characteristic bog shrub is leather-leaf, which grows in profusion and is covered with bell-shaped white flowers in May.

Several rare sedges grow on wet mossy hummocks in Kent Bog, and the showiest is tawny cotton-grass. Long, plumelike bristles extend from their seeds, forming billowy cotton candylike tufts. Much less obvious and rarer yet is the spindly few-seeded sedge, only known in a few Ohio bogs.

Unusual moths often occur in kettle bogs, and the rarest known at Kent Bog is the graceful underwing. Endangered in Ohio, this night-flier's larvae feed on blueberry leaves. Underwing moths are drab and undistinguished—until startled. When threatened, they quickly flash their underwings, which are normally concealed. Bright orange-yellow patterns explode into view like giant eyes, frightening off predatory birds.

It's interesting to ponder changes in Ohio's environment when ensconced among the tamarack stand at Kent Bog, surrounded by northern plants in this glacial relict habitat. Only ten thousand years ago—an incalculable iota of the earth's 4.5 billion years—much of Ohio looked like this. Boreal forests of tamarack and spruce cloaked quaking peaty soils, and plants that we now consider rare and protect were commonplace. Long-lost animals like mastodon, saber-toothed tiger, and muskox ranged the landscape. At Kent Bog, visitors can glimpse our glacial past.

This site is formally known as the Tom S. Cooperrider Kent Bog State Nature Preserve, in honor of the Kent State University botanist who has played a major role in describing Ohio's flora.

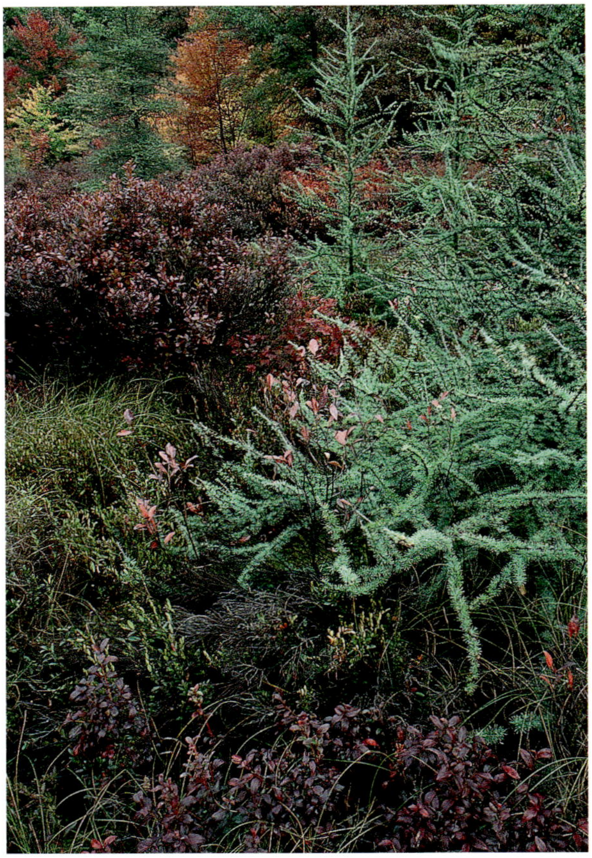

Left: Choked with tamarack and gray birch, Kent Bog contains the largest southern stand of tamarack in the United States.

Opposite: Saddleback caterpillars are small moth larvae that look like they are wearing tiny lime-green horse blankets trimmed in white. Look, but don't touch. Admire them from afar. Stiff clumps of bristles are stinging spines. These hairs can deliver perhaps the most powerful sting of any North American caterpillar.

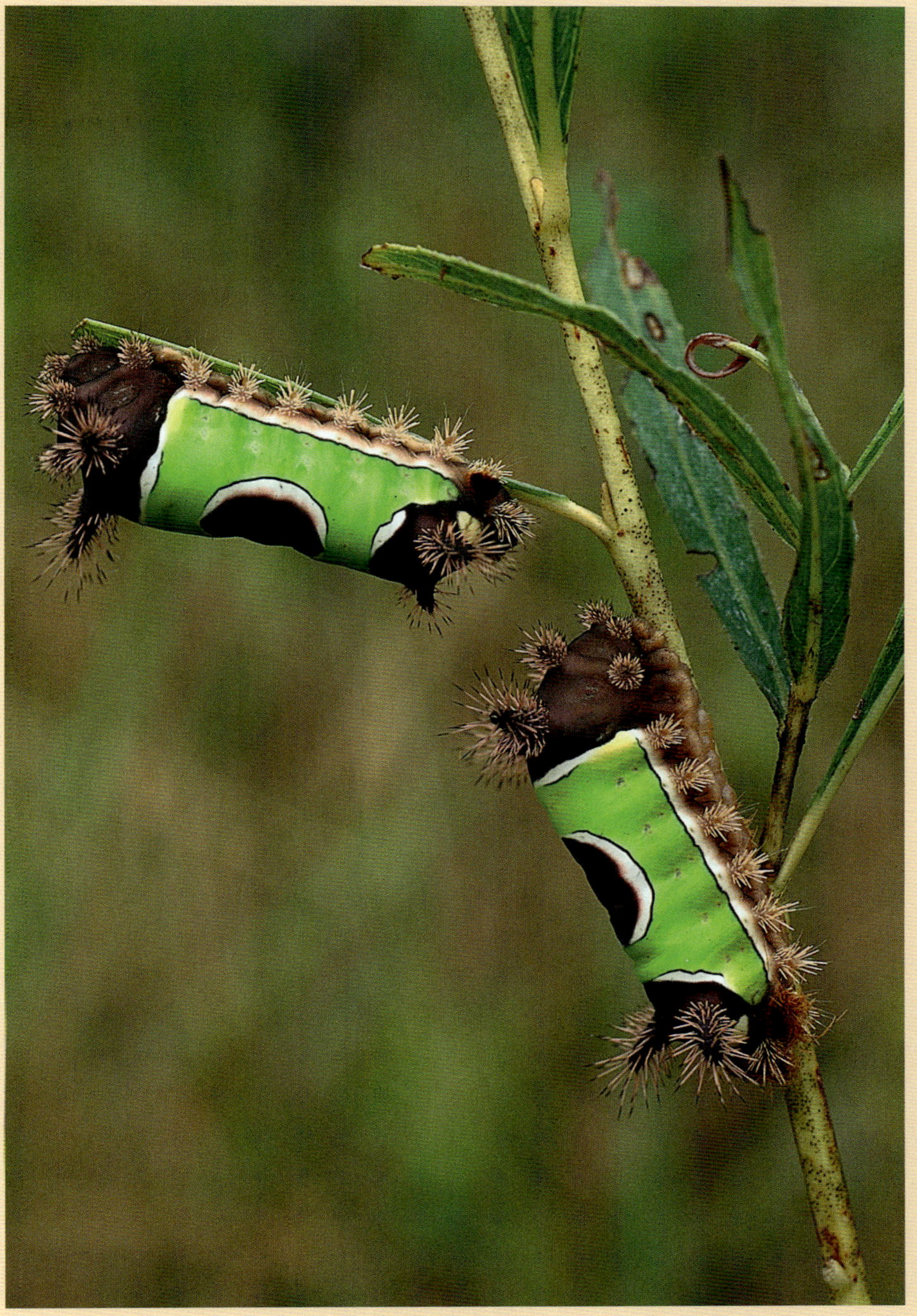

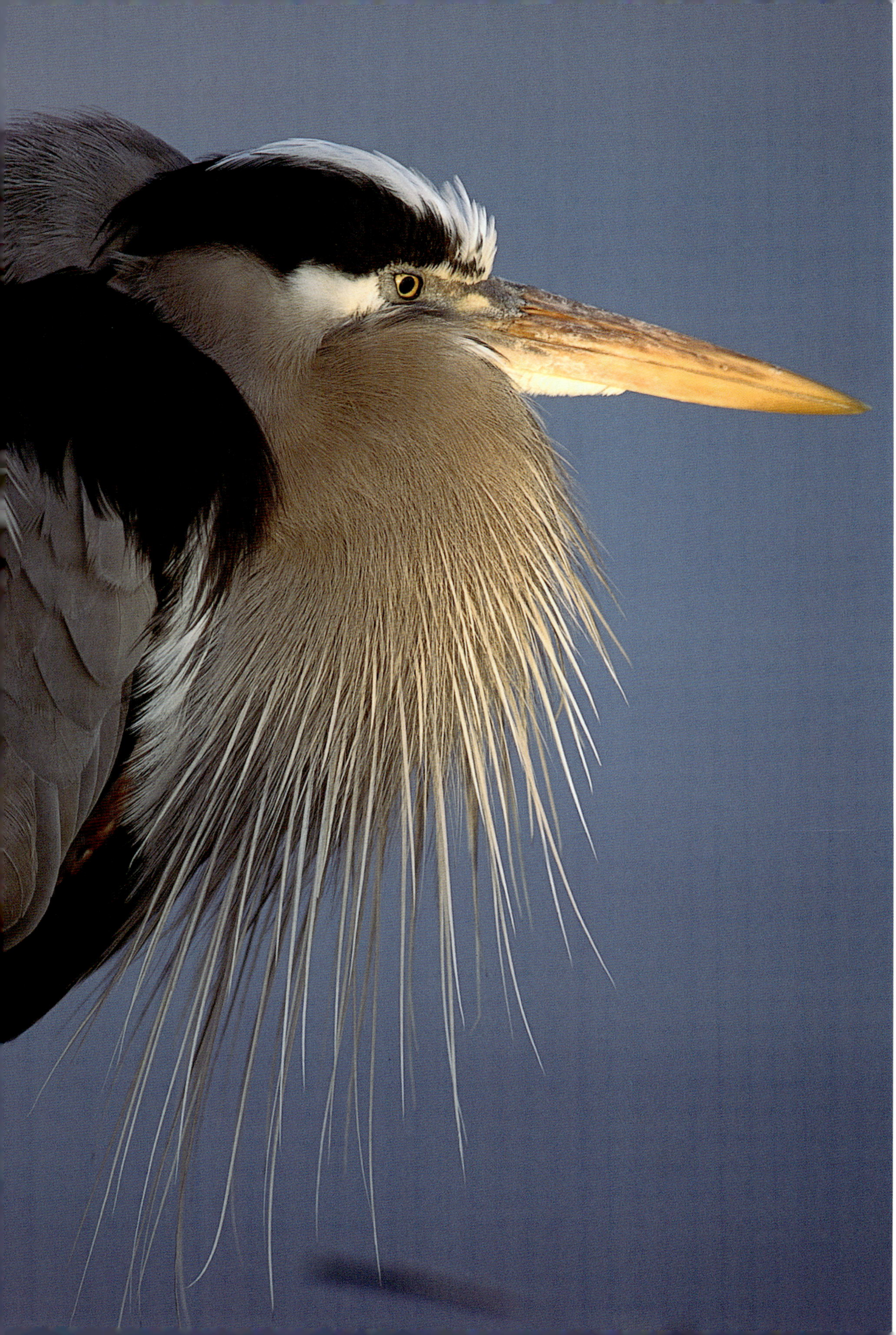

Modern-day floods pale in comparison to the gully-washers of twelve thousand years ago. Meltwaters from enormous ice sheets of the retreating Wisconsin glacier ripped through the landscape in torrents, carving massive channels and creating the vast Killbuck Valley.

The broad, poorly drained valley is filled with wetlands—our largest complex south of Lake Erie. Two types of plant communities dominate: marshes and swamps. Marshes are open wetlands characterized by herbaceous plants like cattails, arrowheads, bulrushes, and water-lilies. Swamps feature woody plants and shrubs, like buttonbush and trees such as green ash, silver maple, swamp white oak, and cottonwood.

Birds rule at Killbuck. Excellent numbers and diversity of waterfowl peak in March, when wetlands are rejuvenated by spring floods. It's possible to find twenty-seven of Ohio's thirty-five species of regularly occurring geese, swans, and ducks—in one day! Hundreds of American wigeon are eagerly picked through by birders seeking the occasional and far rarer Eurasian Wigeon. American wigeon are fun to watch. These colorful dabbling ducks are kleptoparasites; they frequent deeper waters with lesser scaup, redhead, and ring-necked ducks, snatching tasty underwater plants from these diving ducks when they surface.

Four-foot-tall sandhill cranes are on the upswing in Ohio, breeding in perhaps a dozen locales. Until 1987, they hadn't bred in the state

Killbuck Marsh Wildlife Area

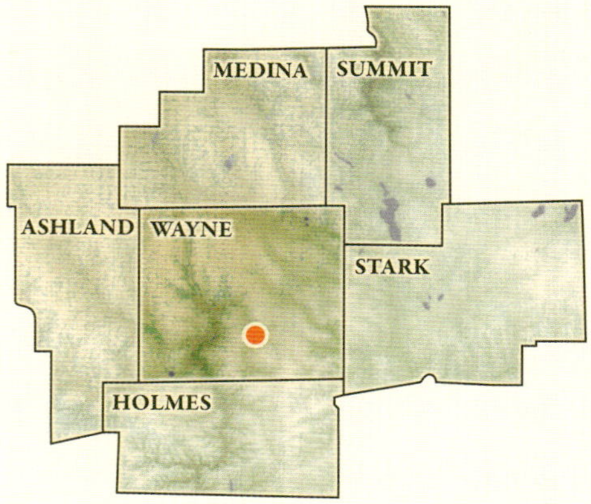

ACREAGE: 5,521

OWNERSHIP: Ohio Division of Wildlife

ACCESS: Open to public during posted hours. Some areas may be access restricted.

WEB SITE RESOURCES: http://www.dnr.state.oh.us/Home/wild_resourcessubhomepage/WildlifeAreaMapsRepository/tabid/10579/Default.aspx

Opposite: A widely recognized symbol of wetlands, great blue herons are common throughout Ohio. Their colonies of nests are known as rookeries and can range in size from a few nests to several hundred.

Right: The only eastern cavity-nesting warbler, prothonotary warblers inhabit swamp forests. Their rich lemon-yellow plumage is practically luminescent. The presence of suitable nesting cavities is a limiting factor in prothonotary warbler distribution. The numerous standing dead trees in the swamps of Killbuck support some of the best populations in Ohio.

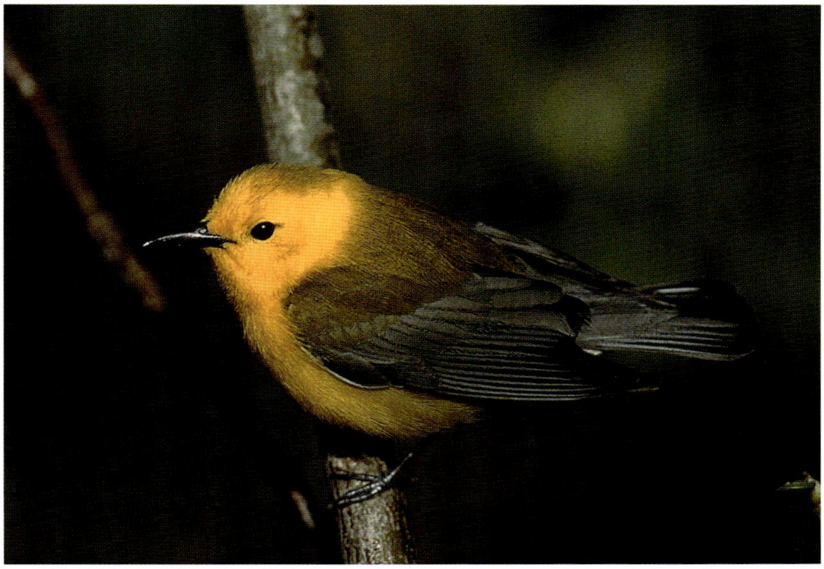

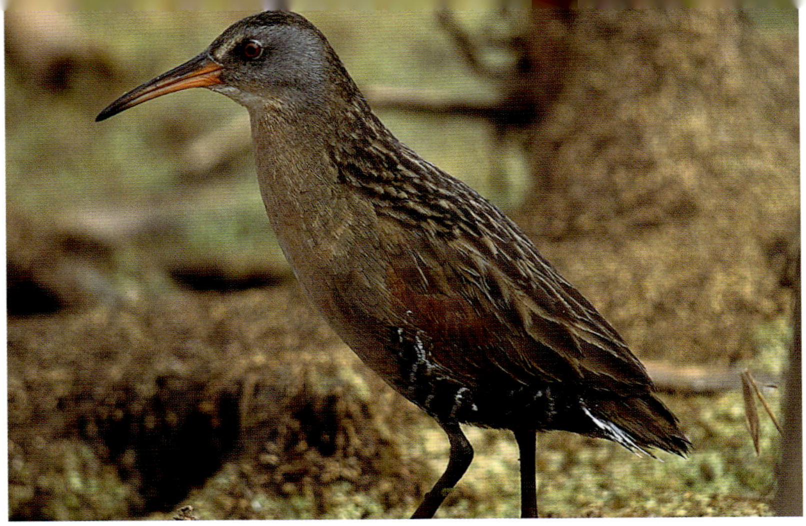

Rails are chickenlike marsh birds with laterally compressed bodies that allow them to easily slip through thick stands of cattails. Virginia rails are far easier to hear than see. They emit a metallic *kiddick kiddick* series of notes.

for nearly one hundred years. That year, a pair successfully nested at Killbuck, and cranes have bred here ever since, expanding to several nesting pairs.

Killbuck's wetlands support many significant breeding birds. Tracts of swampland dead timber are riddled with cavities, and red-headed woodpeckers are frequent. So beautiful are these crimson-headed birds that they inspired the "Father of American Ornithology," Alexander Wilson, to study birds.

Strange sounds emanate from marshy tangles, produced by chickenlike sora and Virginia rails, pied-billed grebes, common moorhens, and least bitterns. Obvious and plucky marshland songsters are marsh wrens. Producing liquid gurgles that sound like a flute being played underwater, marsh wrens are common in cattail and bur-reed colonies. All of these birds have declined due to habitat loss—over 90 percent of Ohio's wetlands have been destroyed.

Aerialists supreme, masses of dragonflies patrol Killbuck wetlands hunting tiny insects. Prominent are skimmers, the largest group of Ohio dragonflies with thirty-seven species thus far recorded. Kaleidoscopelike patterned wings of widow skimmers, twelve-spotted skimmers, and common whitetails shimmer in the sunlight as these insects dart and jag with impossible speed, snapping up gnats and mosquitoes. Jumbo swamp darners also occur; these giants measure three inches and are our largest Ohio dragonfly species.

Wings fill the air at Killbuck, and those with an interest in creatures that fly won't be disappointed.

In 1986, the Ohio Division of Wildlife reintroduced 123 river otters in four locales in eastern Ohio. Their recovery since that time has been swift and successful—an estimated five thousand animals roam the state today, and Killbuck is one of the best places to see one.

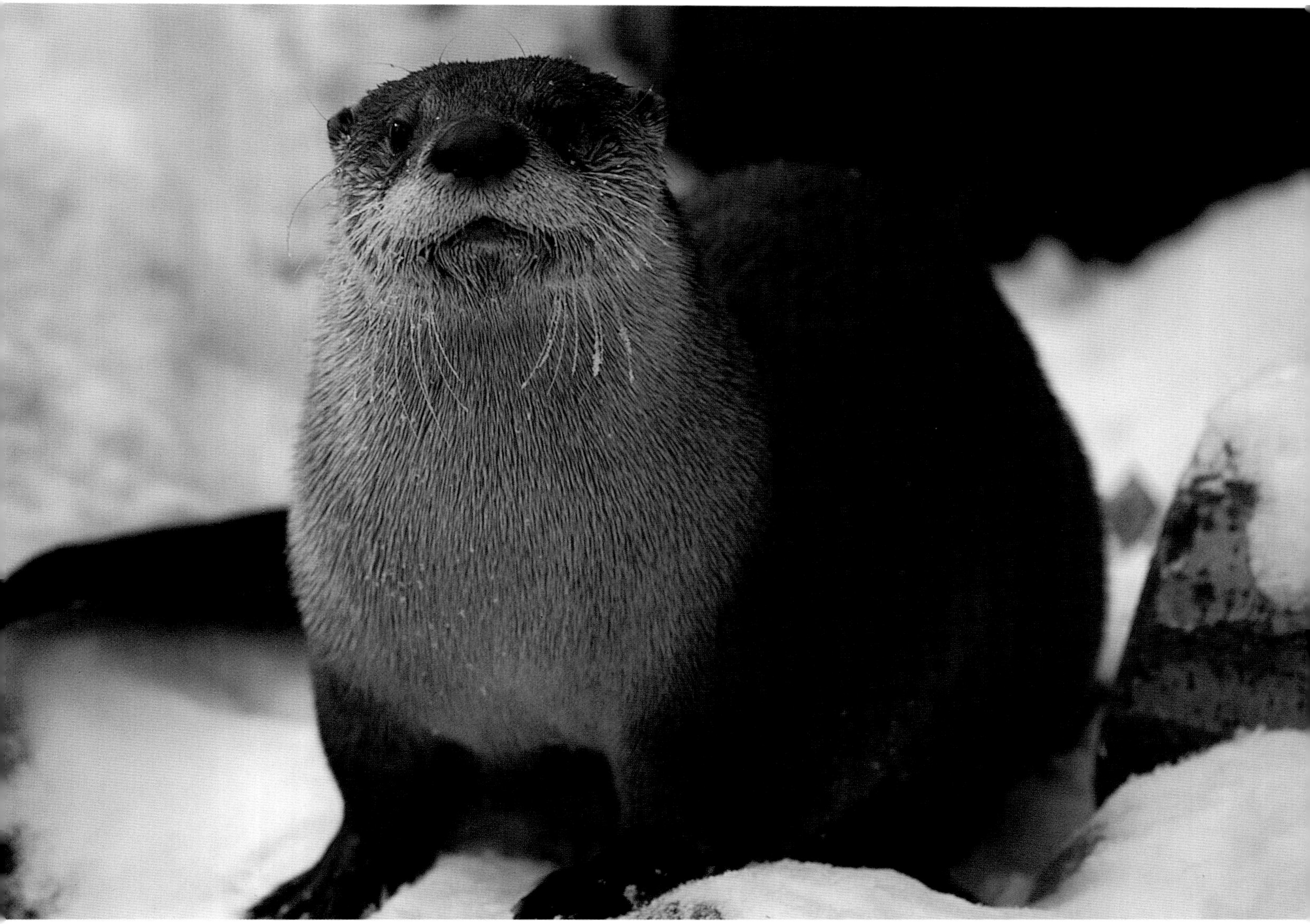

Mantua Bog State Nature Preserve

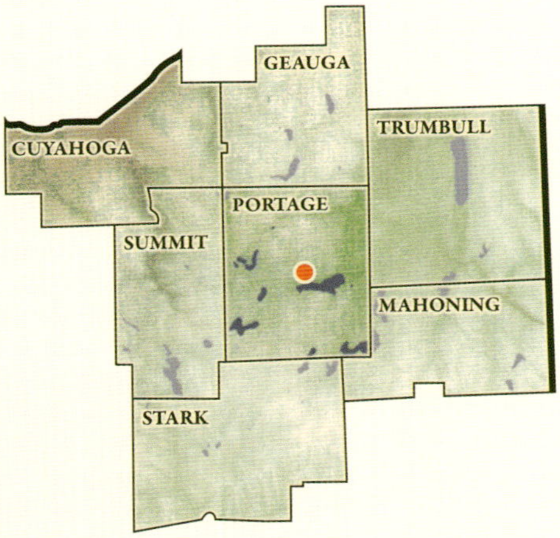

ACREAGE: 44

OWNERSHIP: Ohio Division of Natural Areas and Preserves

ACCESS: By permit only

WEB SITE RESOURCES: http://www.dnr.state.oh.us/dnap/permitonly/tabid/863/Default.aspx

This small preserve is rich in rare flora and fauna. Mantua Bog is a classic example of a boreal fen; many of the plants have strong northern affinities and are at their southern limits here. Although open fen meadows are but a few acres, they are packed with unusual plants. At least twenty-five state-listed rare plants have been found in the preserve.

Thick, spongy hummocks of *Sphagnum* moss create lush specialized substrates for plenty of odd plants. While even hardened botanists are spooked by the daunting prospect of learning sedges, this group of plants is interesting and ecologically important. The genus *Carex* is the largest group in the family, with about 155 species reported in Ohio. Many sedges are habitat-specific and excellent ecological indicator species, like little yellow sedge. Its bristly fruit looks like tiny maces, the medieval weapon used by knights of yore. This rare sedge grows only in fens and other alkaline wetlands. Many other sedges add to the floristic composition of Mantua Bog.

One of Ohio's rarest plants grows in shrubby openings in the fen. Dragon's-mouth is a gorgeous, magenta-flowered beauty and one of North America's showiest orchids. In a good year, perhaps a dozen or two dragon's-mouths appear, and Mantua Bog supports the only known Ohio population. Management activities have helped this orchid to survive, but its existence is tenuous. Although most Ohioans will never see this plant in the wild, at least in their home state, many might be heartened to know that such a beautiful botanical still persists.

Dragonflies abound in Mantua Bog and vicinity, including some rare and local species. By far the most unusual is the brush-tipped

Right: In late summer, the fen meadow at Mantua Bog is covered with ferns, asters, and joe-pye-weed. The reds of poison sumac add to the riot of color.

Opposite top: With brilliant green eyes, state-endangered brush-tipped emeralds can be found at Mantua Bog. Almost all Ohio emeralds are uncommon or rare.

Opposite right: Mantua Bog supports the only known Ohio population of dragon's-mouth orchids. Habitat management at Mantua has helped this orchid to survive.

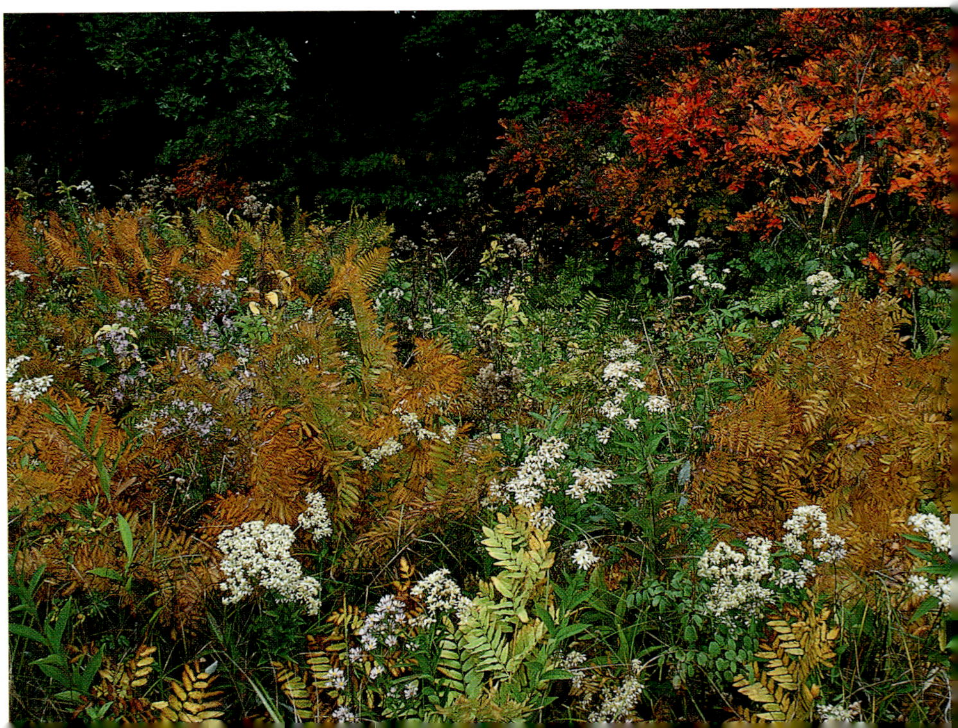

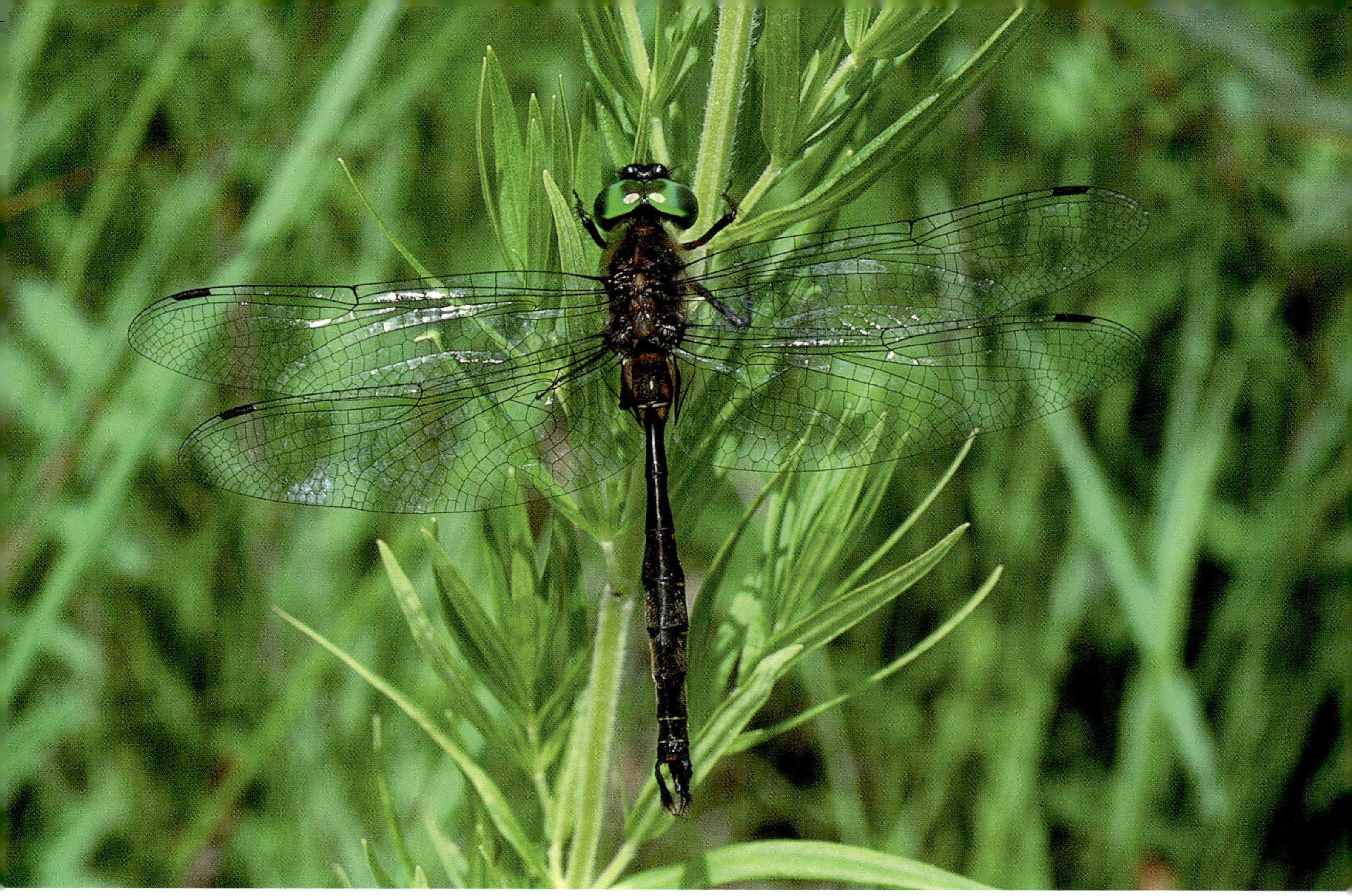

emerald. Each end of this dragonfly is an eye-catcher. The front is capped with two enormous brilliant green eyes, each containing thousands of individual facets. Each facet is the equivalent of a tiny eye; woe to the prey that falls into view of a hungry brush-tipped emerald. The emerald's posterior is adorned with a pair of ferocious-looking hooks, or claspers. These are used to hold the female in place during mating. This endangered dragonfly is only known from Mantua and two other Ohio locales, the southernmost known populations.

Although small, Mantua Bog is significant, a living museum preserving relicts rare or nonexistent elsewhere in Ohio. Protection of sites like this is critical to conserving our state's overall biodiversity.

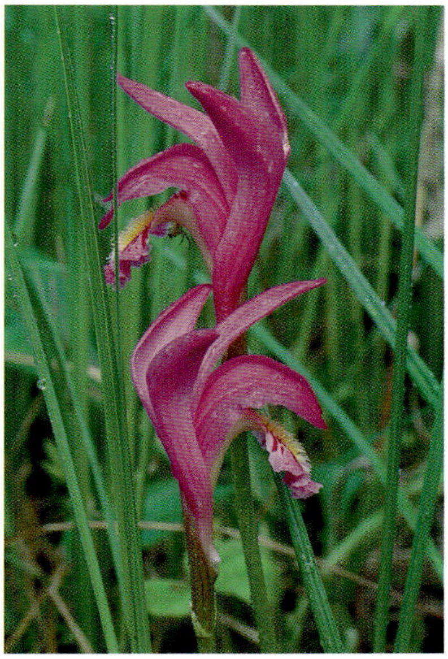

Morgan Swamp

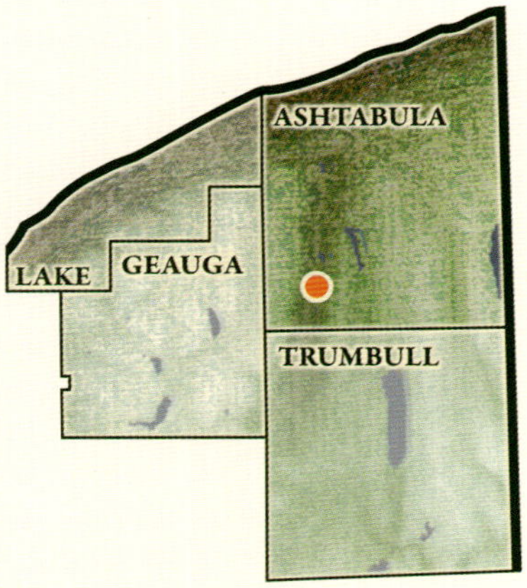

ACREAGE: Approximately 1,000

OWNERSHIP: The Ohio Chapter of The Nature Conservancy

ACCESS: Open to the public during posted hours

WEB SITE RESOURCES: http://www.nature.org/wherewework/northamerica/states/ohio/preserves/art3029.html

Opposite: Spatterdock, marsh fern, and soft rush bedeck a wetland at Morgan Swamp. The dwelling of nature's ecological engineer juts in the background—a beaver lodge. Beaver play a vital role in creating wetlands, and their efforts benefit an enormous diversity of plants and animals.

Gazing upon seemingly pristine Morgan Swamp, it's difficult to imagine the rough history of these tranquil marshes and swamps. The wetlands set on largely impervious clay layers up to fifty feet deep, deposits from twelve millennia ago when this area was covered by a glacial lake. Beginning in the 1850s, loggers began removal of virgin timber that blanketed the swamp, and by 1880 all the trees had been cut. Opening the wetlands to sunlight created drier conditions, and farmers exacerbated drainage efforts in an attempt to grow crops. Fires regularly raged, furthering deteriorating the marsh. Morgan Swamp was described by Lawrence Hicks, one of Ohio's early chroniclers of natural history, as covering about three thousand acres in the early 1930s.

But nature is resilient, and after cessation of logging and farming efforts, Morgan Swamp began to recover. In 1985, the Ohio chapter of The Nature Conservancy began to acquire land here, and has built the preserve into one of Ohio's most expansive privately protected natural areas. Today, the area looks much as it did before European settlement, excepting one sad loss. Before logging, a twelve-hundred-acre primeval hemlock swamp filled much of this low-lying basin, and the giant hemlock were among the first trees harvested. Consequently Ohio lost what may have been our best example of an exceptionally rare habitat.

Beaver have aided greatly in wetland restoration. These large rodents, which can grow to seventy pounds, are tireless workers and brilliant engineers. Relentlessly pursued for their valuable pelts, beaver were extirpated in Ohio by the 1830s. Their reappearance a century later was first recorded in Ashtabula County in 1936. Today they are once again common and creating wetlands. Intricate dams pool water and create a variety of wetland habitats that greatly increase biodiversity.

Morgan Swamp is for the intrepid. The vast, flat terrain pocked with soupy wetlands and tangled vegetation doesn't lend itself to easy navigation, and a compass is recommended. The effort pays off, though—rare plants hard to find elsewhere in Ohio grow on elevated hummocks and in boggy quagmires. At least fifteen state-listed plants have so far been found, and other discoveries await. The rarest is Clinton's wood fern. Found nowhere else in Ohio, this elegant beauty is a northerner, barely nipping into northeast Ohio. There are eighty-five species of ferns documented in the state, and this group is an

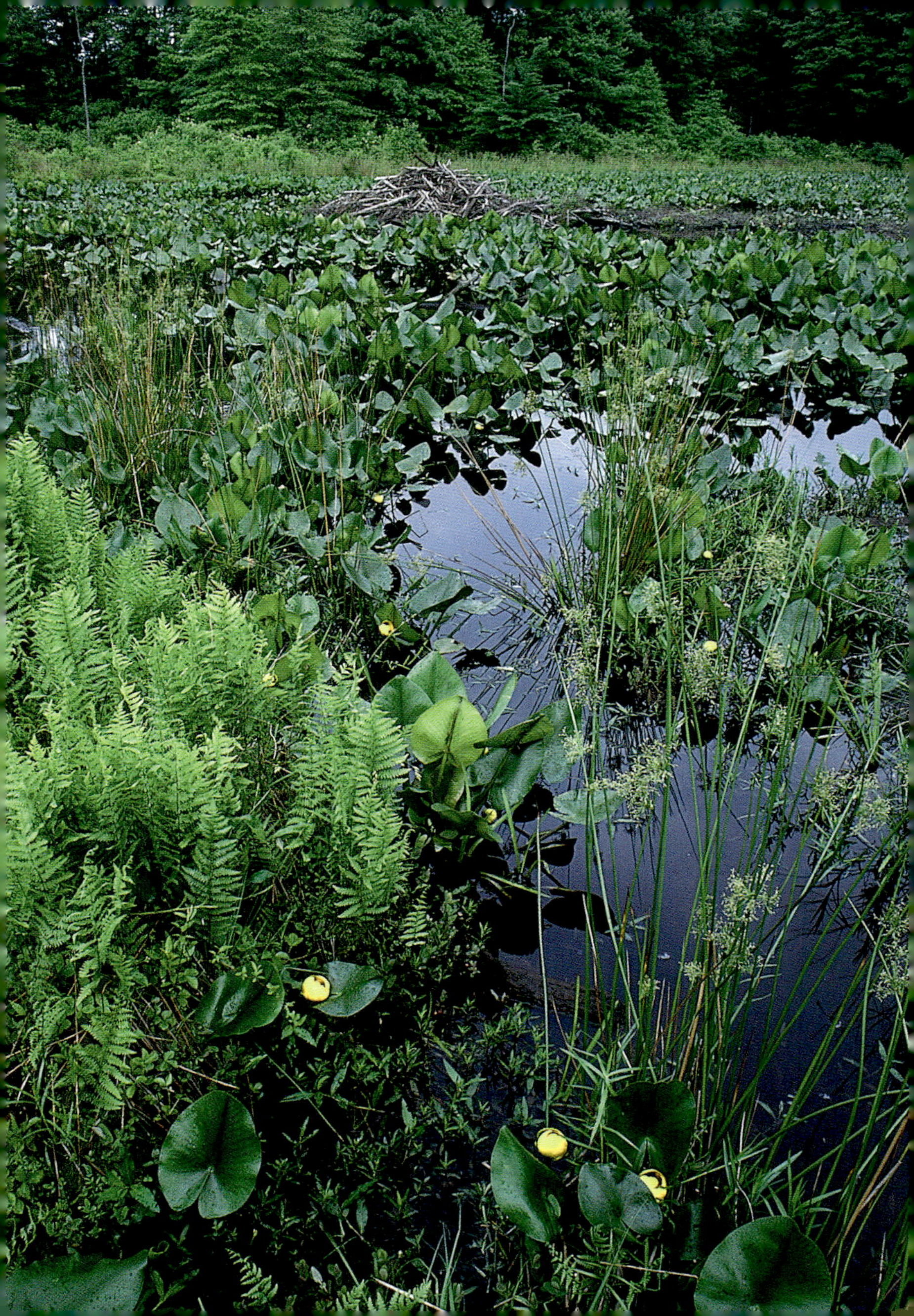

important component of forest ecology. Other plant rarities include wild calla, robin-run-away, and pipsissewa.

Two mammals evince the wildness of Morgan Swamp. There are recent records of both black bear and bobcat, former symbols of Ohio wilderness that are now rare. The former are on the increase, especially in this area where émigrés from expanding Pennsylvania populations turn up. The latter, our only extant wild cat, is more common than thought. Little felines—a big male is forty pounds—they are secretive and more likely to be heard caterwauling in the breeding season than seen.

One of Ohio's last remaining wild streams flows through Morgan Swamp: the Grand River, which ultimately flows into Lake Erie. Dedicated as a scenic and wild river in 1974, its clean waters harbor exceptional aquatic life. Several rare animals sensitive to water-quality degradation survive in this stretch of river. Oddest of all may be the endangered Northern brook lamprey. Primal snakelike fishes, six native lampreys occur in Ohio streams; none are common. Adult brook

Bottom: Many of the bogs and fens in which the spotted turtle occurred have been destroyed. Always hard to find, they forage under the bog mat.

Opposite top: Star-nosed moles are almost completely subterranean and voraciously pursue small prey to fuel their unquenchable metabolisms. Like many moles, they are superb swimmers and often occur in wetland habitats.

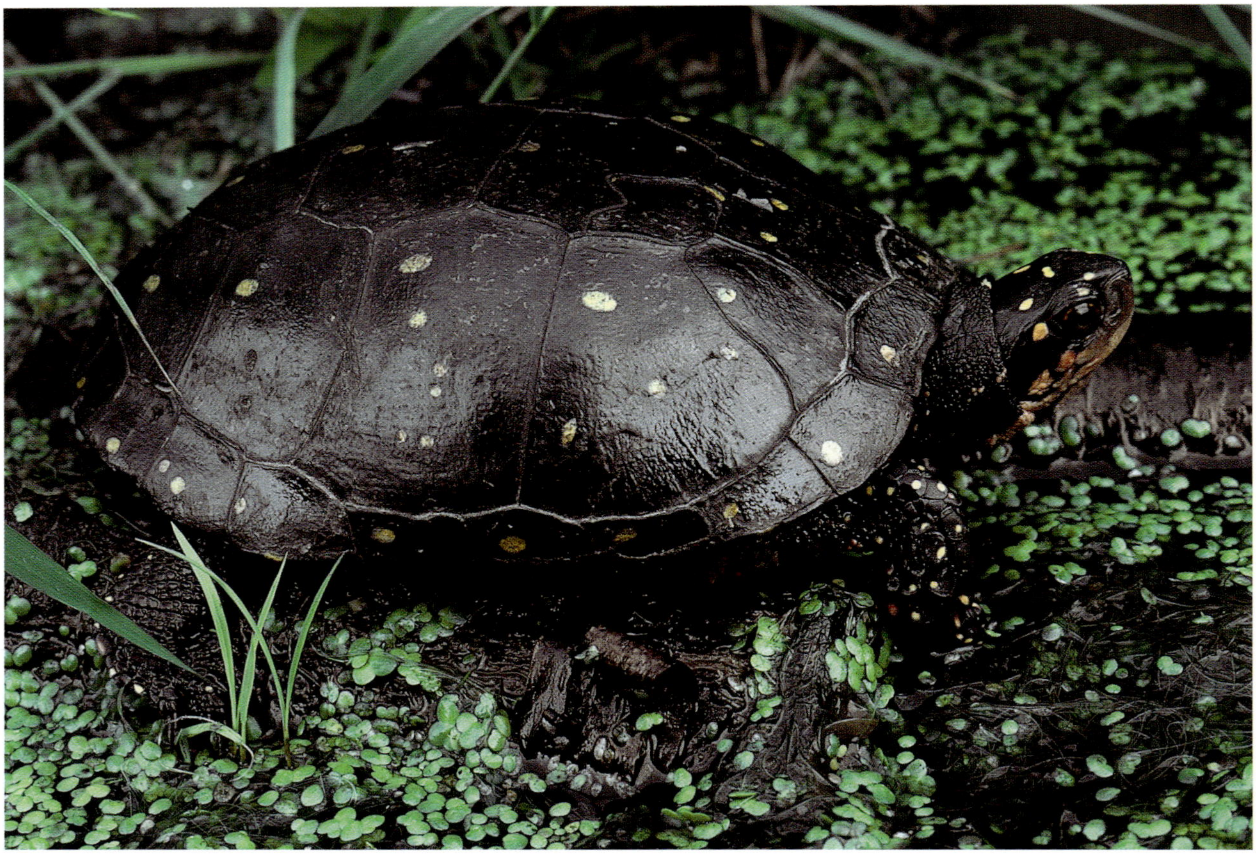

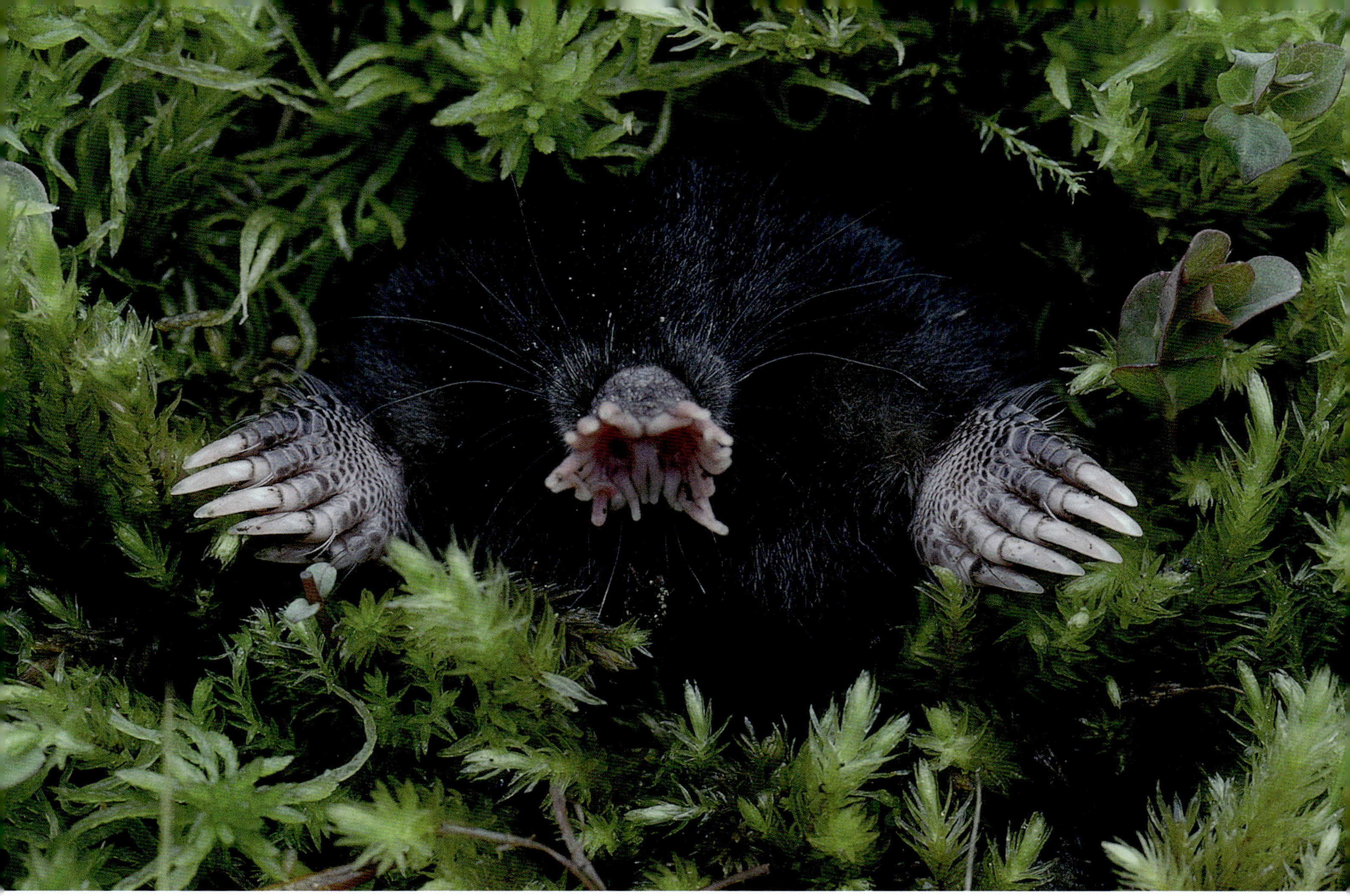

lampreys don't feed; they exist only to mate and reproduce. Other species are parasitic, attaching to host fish via suckerlike mouths by which they grind openings and extract fluids.

The wild lands of Morgan Swamp play a key role in preserving rare biodiversity of northeastern Ohio, including flora and fauna seldom encountered elsewhere. Protection of this area also plays into the big picture of preservation of the Grand River watershed, one of Ohio's last great places.

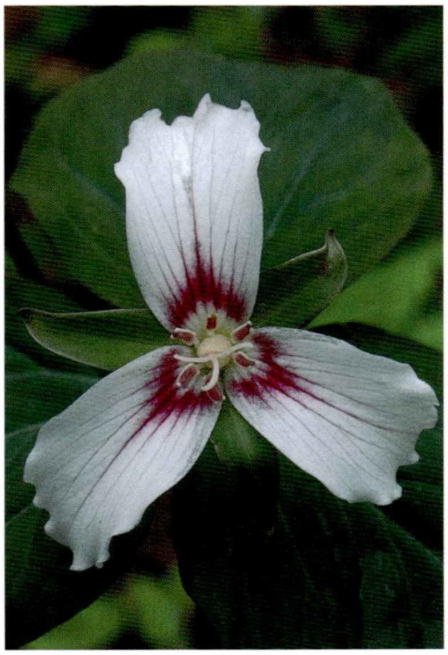

The crimson bull's-eye center of the crisp white flower renders painted trillium unmistakable. Don't expect to see this dazzler in Ohio. A species of the Appalachians, our showiest trillium barely enters northeastern Ohio.

Singer Lake Bog

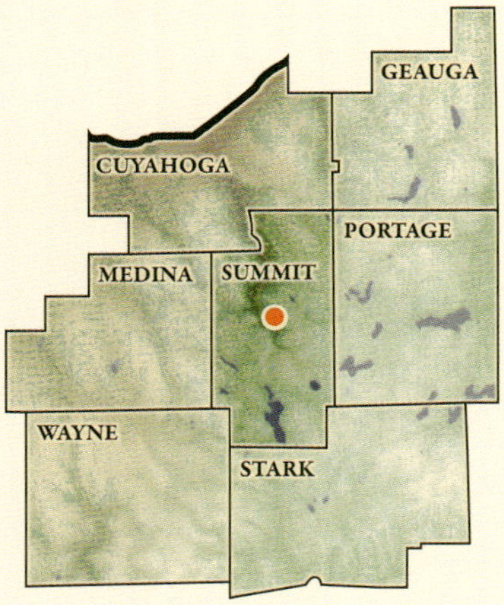

ACREAGE: 344

OWNERSHIP: Cleveland Museum of Natural History

ACCESS: By permit only

WEB SITE RESOURCES: http://www.cmnh.org/site/Conservation_NaturalAreas_Map_SingerLakeBog.aspx

The tale of Singer Lake is one of Ohio's great stories of preservation. Among the state's most significant wetlands, this bog went largely unexplored until the late 1980s despite being in a populous region. Staff of the Cleveland Museum of Natural History made their inaugural visit in 1989. That day, they found the first Ohio record of variegated orange moth. A day-flier, this gorgeous insect depends on cranberry as a host plant—a rare species in its own right. This discovery opened the floodgates of amazing new finds. About forty state-listed rare plants have been documented to date, along with several rare animals.

The bog is vast and intimidating. As with any good peatland, getting around isn't easy, and the careless may drop through the quaking *Sphagnum* mat like an opened trapdoor. Acres of large cranberry carpet spongy ground, and this wetland supports Ohio's largest stand of leather-leaf, a small bog shrub.

On June 1, 2000, intrepid entomologists found two rare dragonflies at Singer Lake: elfin skimmer and chalk-fronted corporal. The tiny skimmer was known only from two other locales. Also endangered, the corporal find was even more incredible. In all, sixty-seven species of dragonflies and damselflies have been found at Singer Lake—perhaps the richest diversity in Ohio, proof of the outstanding quality of these wetlands.

Aquatic plant communities further attest to the pristine nature of Singer Lake. Two endangered species, grass-like pondweed and spotted pondweed, grow in the lake's deep waters. Pondweeds play

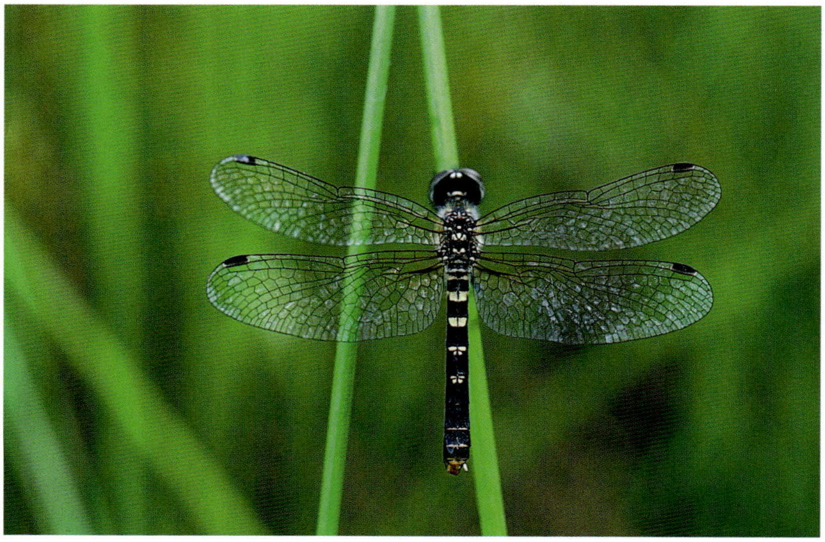

Right: The elfin skimmer, the smallest dragonfly in North America, is endangered in Ohio. The females resemble large hornets.

Opposite: Diverse and expansive, Singer Lake Bog harbors more rare species than any other Ohio bog. Finding unusual flora and fauna isn't easy here; one misstep and an explorer might find themselves neck deep in *Sphagnum* peat.

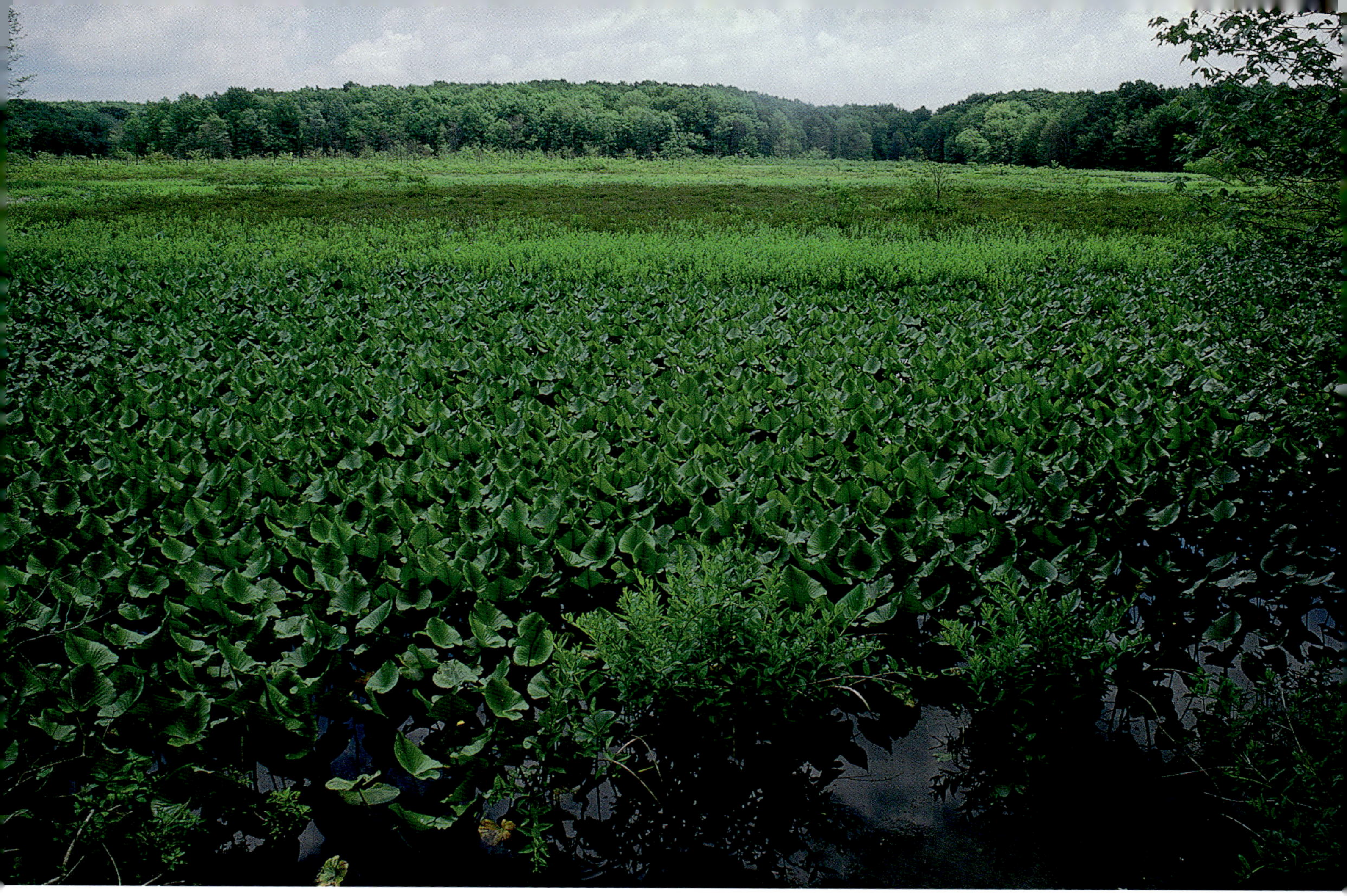

important ecological roles: fish hide in their dense tangles, insects utilize them for food and shelter, and wetland birds eat the kernellike fruit. They are sensitive ecological indicators, and most have declined alarmingly in Ohio. These imperiled plants require clear waters to photosynthesize; murkiness caused by excessive sediments quickly eliminates them. No group of water plants has been harder hit by pollution. Of the twenty-three pondweed species recorded in Ohio, ten are state-listed rarities and five are extirpated.

Few intact relicts of Ohio's postglacial peatland heritage remain. A well-researched 1992 paper published in the *Ohio Journal of Science* documented the loss of 98 percent of the bogs and fens that were present at the time of European settlement. The primary cause of destruction was draining and development. The Natural Areas program of the Cleveland Museum of Natural History worked diligently to save this exceptional peatland. Its labors were successful, and virtually the entire ecosystem is now under protection.

Overleaf top: An unusual member of the gentian family, spring-blooming buckbean occurs in a handful of bogs.

Overleaf bottom: A northern species, the racket-tailed emerald inhabits bogs and fens.

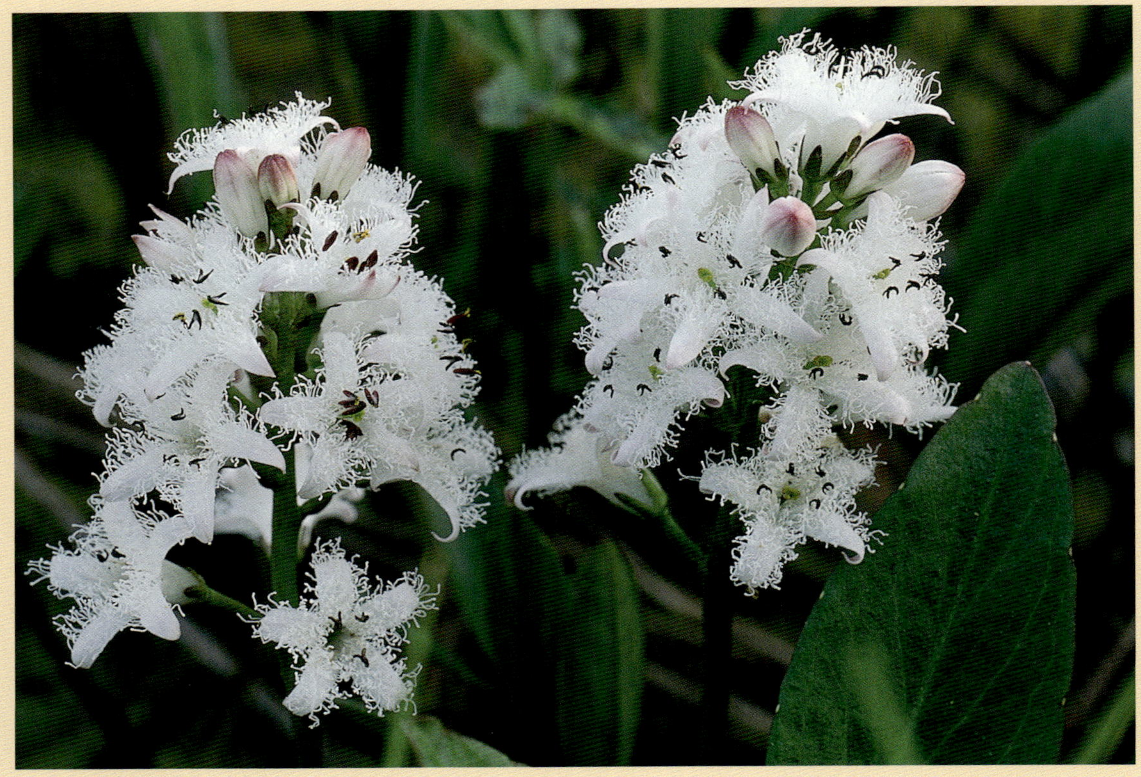

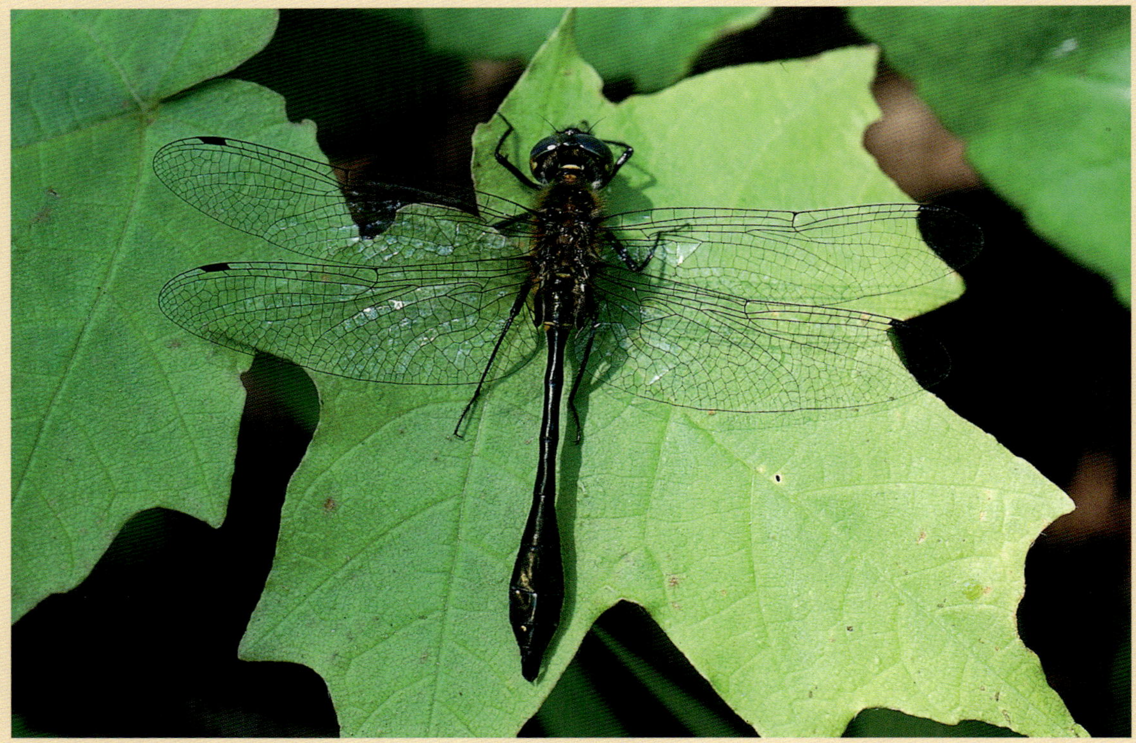

About twelve millennia ago, massive ice sheets of the Wisconsin glacier were in full retreat, melting back to the north. At its southernmost extent, ice covered all but the rough hill country of southeastern Ohio. At the glacier's southern face, occasional gargantuan blocks of ice calved from the sheer frozen walls, leaving great depressions. Water gradually replaced the melting ice, forming a small lake. At first the water was clean and cold and largely barren of life. In time, specialized flora colonized the lakeshores and eventually expanded into the water. Today, most of Ohio's kettle bogs have long since vanished—filled in by vegetation.

Triangle Lake, because of its depth, is still relatively young in the cycle of a bog and offers a snapshot of Ohio's kettle-hole history. Classic bog plant patterns can be studied here. The outer reaches of the kettle are swamp forest, with many red maples. Inside of that zone is tamarack, a deciduous conifer intimately associated with peat bogs. Closer to the lake are shrub zones filled with highbush blueberry, poison sumac, catberry, and other woody plants. Finally, ringing the open lake waters is an open bog mat, composed of various *Sphagnum* mosses.

Bog mats are wonderful, treacherous places. A misstep can land one neck-deep in water, or worse. The risk is worth it to adventurous botanists. Many interesting plants grow on the mat, such as the carnivorous pitcher-plants and sundews. Several bladderwort species grow in little pools amongst the moss. Also carnivorous, these tiny

Triangle Lake Bog State Nature Preserve

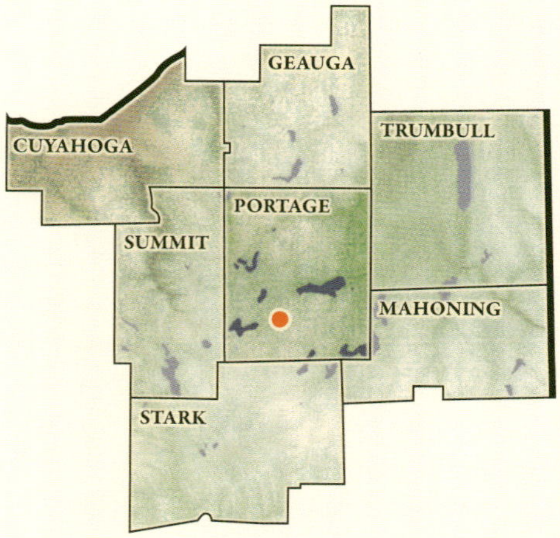

ACREAGE: 61

OWNERSHIP: Ohio Division of Natural Areas and Preserves

ACCESS: Open to the public during posted hours

WEB SITE RESOURCES: http://www.ohiodnr.com/dnap/location/triangle/tabid/967/Default.aspx

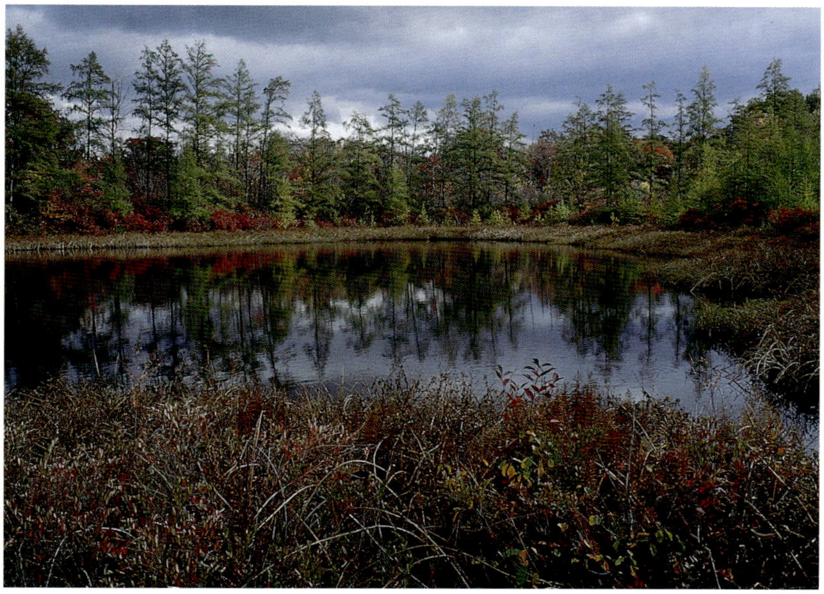

Triangle Lake Bog is Ohio's finest example of a glacial kettle hole. It is an excellent place to learn about the fascinating biology of bogs.

Right: Shown here are two botanical carnivores, pitcher-plant and round-leaved sundew. These plants have evolved highly modified leaves that enable them to capture insects, which serve as their source of proteins and nitrogen.

Below: Masked shrews, Ohio's second smallest mammal, inhabit wetlands. Their bite injects mild neurotoxins to subdue prey.

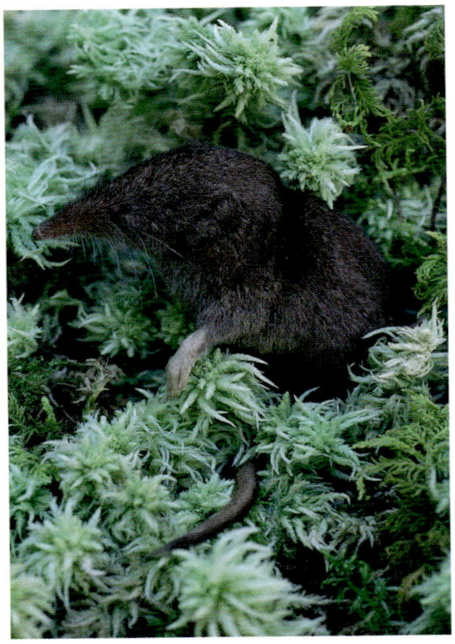

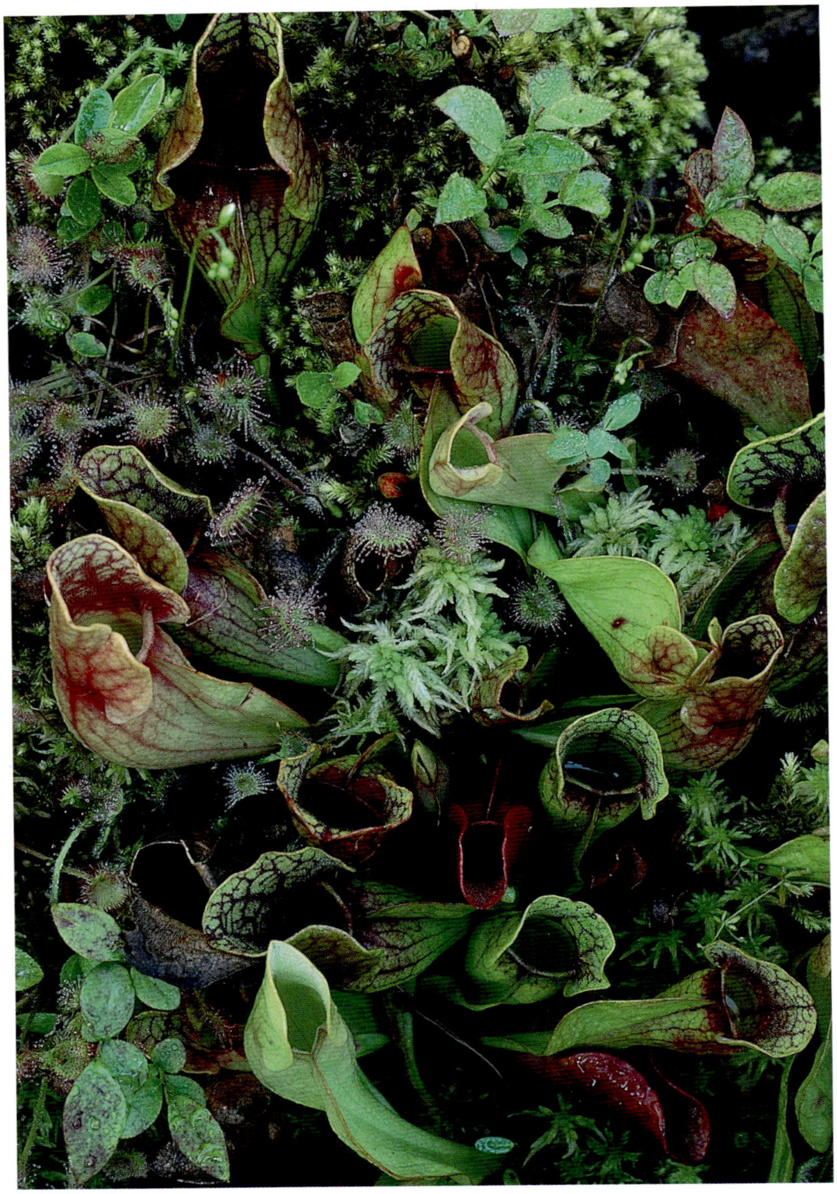

aquatic plants have roots covered with saclike bladders that serve as killing traps, sucking in and ingesting tiny animals.

Triangle Lake, Ohio's finest kettle lake bog, supports an outstanding diversity of damselflies and dragonflies, many of which are very rare this far south. Among them are Hagen's bluet and frosted whiteface. These rarities require pristine waters and specialized habitats, and few such sites remain in Ohio.

Selected Bibliography

Braun, Emma Lucy. *The Monocotyledoneae of Ohio: Cattails to Orchids.* Columbus: The Ohio State University Press, 1967.

———. *The Woody Plants of Ohio.* Columbus: The Ohio State University Press, 1961.

Camp, Mark J. *Roadside Geology of Ohio.* Missoula, Montana: Mountain Press Publishing Company, 2006.

Cooperrider, Tom S. *The Dicotyledoneae of Ohio.* Part 2, *Linaceae through Campanulaceae.* Columbus: The Ohio State University Press, 1995.

Cooperrider, Tom S., Allison W. Cusick, and John T. Kartesz. *Seventh Catalog of the Vascular Plants of Ohio.* Columbus: The Ohio State University Press, 2001.

Daniels, Jaret C. *Butterflies of Ohio.* Cambridge, Minnesota: Adventure Publications, 2004.

Fisher, T. Richard. *The Dicotyledoneae of Ohio.* Part 3, *Asteraceae.* Columbus: The Ohio State University Press, 1988.

Glotzhober, Robert G., and David McShaffrey, eds. *The Dragonflies and Damselflies of Ohio.* Columbus: Ohio Biological Survey, 2002.

Gottschang, Jack L. *A Guide to the Mammals of Ohio.* Columbus: The Ohio State University Press, 1981.

Kaufman, Kenn. *Kaufman Field Guide to Insects of North America.* Boston: Houghton Mifflin Company, 2007.

Lafferty, Michael B., ed. *Ohio's Natural Heritage.* Columbus: The Ohio Academy of Science, 1979.

McCormac, James S., and Gregory Kennedy. *Birds of Ohio.* Edmonton, Alberta, Canada: Lone Pine Press, 2004.

National Audubon Society Field Guide to Wildflowers (Eastern Region). 2d ed. New York: Alfred A. Knopf, 2001.

Ohio Division of Wildlife. *Amphibians of Ohio.* Publication 348. Columbus, Ohio, 2006.

———. *Reptiles of Ohio.* Publication 354. Columbus, Ohio, 2006.

Peterjohn, Bruce G. *The Birds of Ohio.* 2d ed. Wooster, Ohio: Wooster Book Company, 2000.

Platt, Carolyn, and Gary Meszaros. *Birds of the Lake Erie Region.* Kent, Ohio: Kent State University Press, 2001.

———. *Creatures of Change: An Album of Ohio Animals.* Kent, Ohio: Kent State University Press, 1998.

Ramey, Ralph. *Fifty Hikes in Ohio.* 3d ed. Woodstock, Vermont: The Countryman Press, 2007.

———. *Fifty More Hikes in Ohio.* Woodstock, Vermont: Backcountry Guides, 2002.

Rosche, Larry. *Dragonflies and Damselflies of Northeastern Ohio.* Cleveland, Ohio: Cleveland Museum of Natural History, 2008.

Showman, Ray, and Don Flenniken. *The Macrolichens of Ohio.* Columbus, Ohio: Ohio Biological Survey, 2004.

Trautman, Milton B. *Fishes of Ohio.* Columbus: The Ohio State University Press, 1957.

Index of Animals and Plants

Fishes

Freshwater Mussels

Insects

Lichens

Mammals

PLANTS

Reptiles